structural
detailing
for technicians

structural detailing
for technicians

Gerald L. Weaver

Architect
Dodge City, Kansas

McGraw-Hill Book Company

New York
St. Louis
San Francisco
Düsseldorf
Johannesburg

Kuala Lumpur
London
Mexico
Montreal
New Delhi

Panama
Rio de Janeiro
Singapore
Sydney
Toronto

Library of Congress Cataloging in Publication Data

Weaver, Gerald L
  Structural detailing for technicians.

  Bibliography: p.
  1. Structural drawing. I. Title.
T355.W34   604'.2   73-21899
ISBN 0-07-068712-9

STRUCTURAL DETAILING FOR TECHNICIANS

10 11 12 13 14 15 EBEB 89 88 87

The editors for this book were
Robert Buchanan and Myrna Breskin,
the designer was Marsha Cohen, and its pro-
duction was supervised by Patricia Ollague.
It was set in Trade Gothic by Progressive
Typographers.
It was printed and bound by Edwards Broth-
ers Incorporated.

contents

Many drafting and engineering drawing texts include a chapter devoted to structural draw-ing, but relatively few have been devoted exclusively to the subject. The number of struc-tural detailing textbooks aimed at the specific needs of technicians is even more severely limited. This book has been written in an effort to provide more adequate coverage of the structural detailing process and to prepare the student to produce basic structural drawings for use in manufacturing, construction, engineering, and architectural firms. I hope that it will also prove to be a useful source of review and reference for draftsmen and technicians already working in the industry.

This text provides the basis for a semester-length course in structural detailing at the tech-nical school or junior college level. A desirable prerequisite for the course would be at least one basic drafting or engineering drawing course.

The first three units of the text cover structural systems utilizing steel, wood, and concrete, with emphasis on the composition and characteristics of the material itself. The fourth unit presents an overview of the emerging technological advancements destined to become more common in structural systems of the future.

preface

The text material and the exercises at the end of each chapter are intended to portray the actual problems and practices prevalent in industry today. The text intentionally omits the area of structural design, which is the function of the professional engineer or architect. Any work written on structural detailing must rely heavily on the many standards and man-uals currently used by detailers. Portions of these handbooks of standard components and practices are quoted throughout the text, and acknowledgments appear with the material as it occurs.

I am appreciative of the assistance given me by the associations and professional prac-titioners of the structural detailing art. Their comments have helped to ensure the adequacy and accuracy of the material.

If students sincerely apply themselves to the subject and satisfactorily complete the as-signed work, they will find a place as a member of that very necessary team of specialists responsible for construction of every type. It is my hope that this volume will play an im-portant role in achieving that goal.

Gerald L. Weaver

to the student

This text is meant to be used as an instructional resource to help you acquire a basic understanding of structural detailing. Hopefully, it will serve its purpose well.

Since I feel it will be advantageous for you to know ahead of time what information you should be looking for, this text has a study guide at the beginning of each chapter.

A course in structural drafting should develop both a knowledge of the subject and the skills necessary to succeed at the on-the-board drafting process. The problems assigned at the end of each chapter were selected to provide realistic practice and experience in structural detailing. I heartily recommend that you complete each problem to the best of your ability.

The examples shown throughout the text are provided for assistance in understanding structural detailing terminology and practices. Although they are realistic and representative, they should not be used to construct actual structures or be relied upon as significant design data.

The first faltering attempts to use iron for structural members resulted in columns and beams which were bulky, unsightly, and often unsafe. The problem of adequately fastening several iron components together constantly plagued early builders. Designers developed more finesse when it became economically feasible to use the material in the form of cast iron.

It appears that the structural use of cast iron was tied closely to the development of railways. Railroads demanded bridges with long-span capability, and railroad stations called for the enclosure of large, open spaces (particularly train sheds and shops) with long-span roof systems.

An iron bridge at Coalbrookdale, England, constructed in 1777, was one of the first efforts to rely heavily upon iron for the structural system. The technology matured rapidly during the ensuing years, and cast-iron pillars and columns were being used in mill construction by 1800. The later part of the Early Victorian period in England found cast iron in frequent use in conjunction with wrought iron at points where tension stresses were anticipated. The most common early use of iron structural members was in combination with other materials, where the iron could be hidden from view.

Exhibition halls provided designers with the opportunity to experiment with structural cast iron in an honest, exposed manner. The cast-iron members were combined with large areas of glass for totally new effects. The Crystal Palace, designed by Sir Joseph Paxton, was built to house an exhibition in London in 1850. The structure relied upon a cast-iron structural system, with roof bays enclosed with glass between the structural members. The total width of the building was 408 ft, but the use of multiple bays cut the span down to 72 ft. Total length of the Crystal Palace was 1,848 ft, and the center height was 72 ft. The potential of cast iron for prefabrication was clearly demonstrated when this immense structure, enclosing some 750,000 sq. ft, was fabricated and erected in approximately nine months.

The Eiffel Tower in Paris is a familiar example of the early development of iron structural systems. This remarkable structure, constructed in 1887, soars to a height of 984 ft. Gustave Eiffel, the engineer responsible for the design, had created a number of smaller structures utilizing iron in the same fashion.

Cast iron used for numerous projects in the mid-1850s was often unreliable because of the methods employed in its manufacture. The material was also extremely vulnerable to damage by fire. Early engineers had hoped that iron structural members would produce fireproof buildings and allow the light structural system to be exposed. A series of fires destroyed those hopes and many fine examples of the emerging technology. Designers were forced to protect iron members with masonry, and beams and columns were wrapped with brick as fireproofing.

The mid-nineteenth century witnessed development of the iron-rolling process and the initial use of rolled-iron floor beams. Upper floors were constructed on the iron beams, using brick or structural tile arches (and later reinforced concrete) for the floor system. Exterior walls were load-bearing, and cast-iron columns served as interior supports for floor and roof beams.

Original attempts to construct high multistory buildings were hampered by the extreme thickness of walls at the base necessary to support the weight of the wall above. Construction costs of such buildings were prohibitive, but rising land prices in urban areas of the United States demanded multistory structures. Invention of a workable elevator further prompted the need for, and feasibility of, taller buildings. Engineers were forced to reassess their efforts and strive to develop a more satisfactory structural system.

Through experimentation, it became possible to achieve consistent manufacturing results with metal alloys. An alloy is simply a mixture of two or more metals, and such a mixture offers the potential of greater strength, greater durability, more satisfactory working qualities, and decreased component size and weight.

The satisfactory production of the alloy known as steel, around 1870, gave the necessary qualities to iron to enable the technology to advance. Engineers, having experienced success with rolled-iron beam systems, discovered that exterior wall thicknesses could be greatly decreased and supported on horizontal steel beams (as they had supported floor weight in the past). These factors allowed structural systems to advance steadily to their present state.

COMPOSITION OF STEEL

Metals can be classified into two distinct groups. Metals with iron content are categorized as ferrous metals, while those containing no iron are called nonferrous metals. The ferrous metals include iron, cast iron, steel, alloy steel, and galvanized iron. Examples of nonferrous metals are aluminum, brass, copper, lead, gold, silver, and platinum.

A number of metals (both ferrous and nonferrous) are currently used in the construction industry, but the most prevalent structural metal is steel. Steel is an alloy using iron as its base. Structural steel is generally of the carbon-steel family and contains varying amounts of carbon, depending upon the purpose of the structural part, or member.

The strength and working qualities of steel can be altered considerably by the many fabrication and treatment processes available. The same basic material may offer many different characteristics when stamped, cold-rolled, hot-rolled, extruded, or cast.

STEEL MANUFACTURE

Three basic methods are primarily used in producing steel, and the steel produced by each method has unique characteristics. The basic production processes are the Bessemer, open-hearth, and crucible processes.

Bessemer steel utilizes a process developed by an English engineer, Sir Henry Bessemer. In a large container called a Bessemer converter, molten pig iron is purified by blasting air through the molten material to burn away carbon and impurities. The material is then recarbonized to the proper degree by adding specified amounts of *spiegeleisen* (a hard, white pig iron containing from 15 to 32 percent manganese).

Open-hearth steel gets its name from the low-roofed furnace used in the process, a structure with a wide, saucer-shaped hearth. Pig iron is melted and wrought-iron scrap is added until the proper degree of carbonization has been achieved.

Crucible steel is a high-quality product used for cutting edges and projectiles. The name refers to the covered container (or crucible) used in the process. Impure wrought iron is cut into small pieces and placed in the crucible. Desired amounts of carbon and other alloying elements are placed on top of the wrought iron, and the entire contents of the crucible are melted in the covered container.

STANDARDIZATION

Economy in manufacturing depends heavily upon the standardization of a product to take full advantage of mass-production techniques. Such standardization results in lower costs to the consumer. Structural steel has been standardized to a great degree. A number of governmental agencies and industrial associations have had a hand in setting and maintaining standards for structural steel products.

The American Institute of Steel Construction (AISC) has been, and continues to be, a cohesive force in the standardization of steel products intended for the construction industry. The *Steel Construction Manual* published by the organization is a comprehensive reference work widely used by those involved in the design and construction of steel structures. The manual carefully delineates the full range of structural steel components available as standard production items from steel manufacturers.

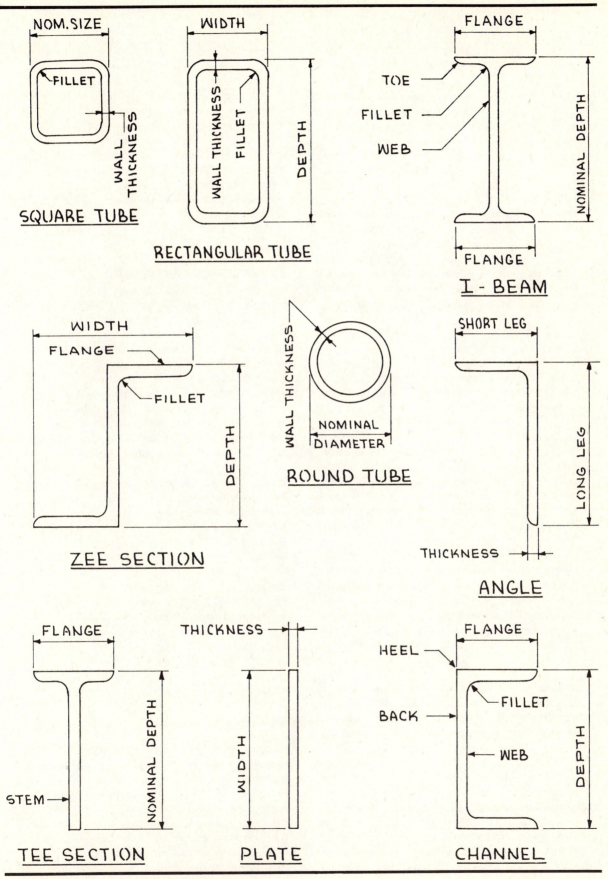

Figure 1·2 Sectional shapes of steel members

STEEL SECTIONS AND COMPONENTS

Since the need for economy has dictated that structural steel be manufactured in stock sizes and shapes, the structural designer calculates the size necessary for each member and chooses the stock size closest to, and larger than, his calculated section. Structural members are designated by their cross-sectional shape (see Figure 1-2). Many columns and beams utilize the I-shape. This section, resembling the capital letter I and often called the I-beam, is much stiffer in resisting bending when the load is applied to the flange area (top) of the I. Light sections of I-beam shape are often called junior (Jr.) beams.

A variation of the standard I-beam has been the wide-flange beam, now designated W (formerly WF). This section has wider flanges than the I-beam in relationship to the web distance and is sometimes referred to as an H-section.

Channels are sections shaped like the letter C. Like the I-beam and H-section, channels are strongest when the load is applied in the direction of the web. Two designations are used for channel sections: C designates American standard channels, and MC denotes miscellaneous channels.

Angles are shaped like a capital L and are available with equal or unequal legs. T-bars are available, but usually perform a subordinate role such as supporting ceilings. They are cut from the standard W-, S-, or M-sections and labeled by section as WT, ST, or MT.

Z-bars are used in applications best suited to their peculiar shape. Plates (flat stock) are widely used for built-up sections and can be fabricated by welding, riveting, or bolting with high-grade steel bolts.

Round tubular sections are often used for columns and can be combined with concrete to form Lally columns. Round, square, and flat bars are standard items with many manufacturers. Prefabricated built-up beams and members are available, and include the much-used bar joist or open-web steel joist.

Table 1-1 gives the following information about the various sections: name, shape, general size range, weight range in pounds per linear foot, and an example of the typical method for calling it out on a drawing. Notice that for most shapes the call-out contains in this order the depth, shape symbol, and weight per foot. This system covers the W-, S-, and M-beams, light columns, and the American standard and miscellaneous channels. Angles are called out by giving the symbol, the dimensions of each leg, and the thickness of the material.

COMPOSITE SECTIONS

The prefabricated joist units shown in Figure 1-3, the bar joist or open-web steel joist, are available in a number of series. The overall cross-sectional shape of the units conforms to the I-shape. Steel angles form the flanges (back to back, with web members between), and the web is made up of round steel bars, undulating to form triangular bracing between the top and bottom chords (flanges). The standard end of such joists allows only the top flanges to run through to the supporting wall, or beam, but these joists are available with the bottom flange or chord extended to permit hanging ceilings adjacent to the wall line. Joist series fall into two main categories: the short-span and the long-span series.

Short-span bar joists are designated in the J-series and are available in a high-strength configuration in the same sizes, termed the H-series. Depth of the J- and H-series ranges from 8 to 24 in., and the weight per linear foot runs from 4.2 to 12.4 pounds. The short-span series is designated by calling out the manufacturer's number, giving depth first: 8 J 2, 10 J 2, 24 J 8, etc.

Long-span bar joists are produced in the regular LA-series and in the LH-series (the high-strength version). Depth of these units ranges from 18 to 48 in., and the weight per linear foot runs from 13 to 68 pounds. Designation follows the pattern for short-span joists; thus call-outs carry such nomenclature as 18 LA 02, 48 LA 19, etc. Some areas delete the A and call out simply 18 L 02, 48 L 19, etc.

TABLE 1-1 STEEL SECTIONS, SIZES, AND NOMENCLATURE TABLE

Name	Shape	General Size Range	General wt. Range lb/lin. ft.	Call-out
Junior beams	I	1 7/8–3 in. width, 6–12 in. depth	4.4–11.8	8 Jr 6.5
Light beams	I	4 in. width, 6–12 in. depth	12–22	8 B 15
American Standard beams	I	2 3/8–7 7/8 in. width, 3–24 in. depth	5.7–120	10 I 35
Wide-flange beams	I	4–16 1/2 in. width, 4–36 in. depth	13–300	10 W 21
Light columns	I	4–8 in. square	13–34.3	6 × 6 M 25
Junior channels	C	1 1/8–1 1/2 in. width, 10–12 in. depth	6.5–10.6	10 Jr U 6.5
American Standard channels	C	1 3/8–4 1/4 in. width, 3–18 in. depth	4.1–58	7 U 9.8
Equal-leg angles	L	1 × 1 in.–8 × 8 in.	0.80–56.9	L 4 × 4 × 3/4
Unequal-leg angles	L	1 3/4 × 1 1/4 in.–9 × 4 in.	1.23–40.8	L 7 × 4 × 7/8
Structural tees (by splitting webs of beams)	T	1.8–16 in. width, 2.9–18.3 in. depth	2.2–150	ST 5 W 10.5 ST 6 I 20.4 ST 6 B 9.5 ST 6 Jr 5.9
Tees (flange by stem)	T	2–3 1/8 in. width, 2–5 in. depth	3.5–13.6	T 3 × 3 × 6.7
Zees	Z	2 11/16–3 5/8 in. width, 3–6 1/8 in. depth	6.7–21.1	Z 4 × 3 × 15.9
Plates, bars, and bearing plates	N.A.	3/16–4 in. thick, 6–186 in. width	Wt. varies	PL 18 × 1
Square bars	□	1/16–12 in. square	0.013–489	Bar 1 ⧄
Round bars	○	1/16–12 in. diameter	0.01–384	Bar 2 1/2 ⌀
Flat bars	▭	3/16–1 in. thick, 1/4–128 in. width	.16–435	Bar 2 1/4 × 1/2
Bearing piles	H	8 × 8 in.–14 × 14 1/2 in.	36–117	12 BP 74
Pipe columns	○	3–12 in. diameter	7.58–72	3 in. pipe Col. (std.) or 3 in. O.D. × 0.250
Square structural tube	□	3 × 3 in.–12 × 12 in.	16–50	4 × 4 RT × .250
Rectangular structural tube	□	3 × 6 in.–8 × 12 in.	29–46	4 × 6 RT × .250

NOTE: Other specialty series are produced, including car and ship channels and bulb angles, but because of the relatively infrequent use in building construction, they are not listed above.

Plate girders are actually oversize built-up beams. Their primary application lies in the area between standard rolled beams and trusses. Plate girders are built up using standard sections (angles, channels, etc.) in conjunction with steel plate. The shape of the finished unit most often resembles the I. Fabrication may involve welding, riveting, or both. The plate girder allows the engineer to design members which are nonstandard, basing them upon combinations of standard components. See Figure 1-4.

Plate and angle columns are often used and are similar to plate girders in design and fabrication. They rely heavily upon the standard sections as their components (Figure 1-4).

W-sections used for columns are sometimes provided with additional plates covering the flange areas. These sections are referred to as cover-plated columns. Other structural combinations are often designed combining channels, beams, angles, etc., to fit specific purposes which single standard components cannot fulfill (Figure 1-4).

Rigid frames as shown in Figure 1-5 are well suited for enclosure of large, open areas. These units are fabricated by a number of manufacturers and, when

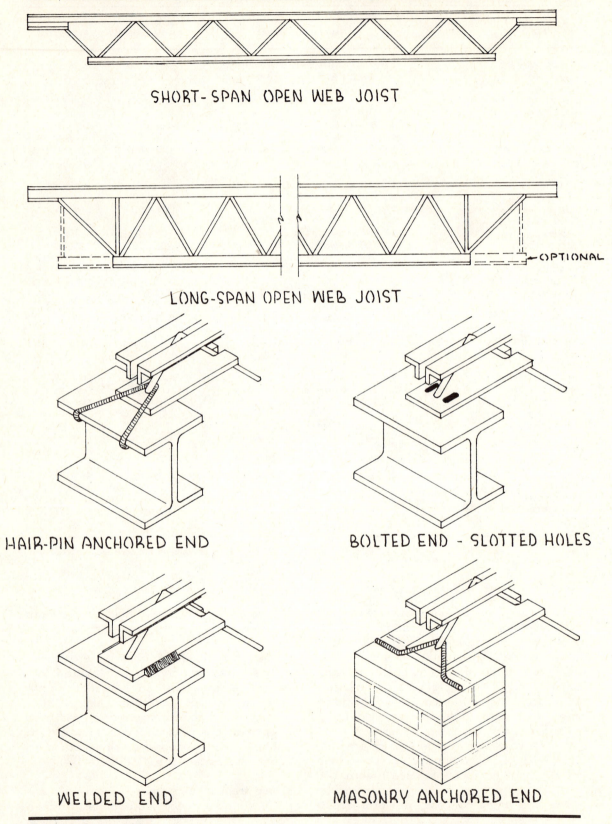

SHORT-SPAN OPEN WEB JOIST

LONG-SPAN OPEN WEB JOIST

←OPTIONAL

HAIR-PIN ANCHORED END

BOLTED END - SLOTTED HOLES

WELDED END

MASONRY ANCHORED END

Figure 1·3 Open web joists

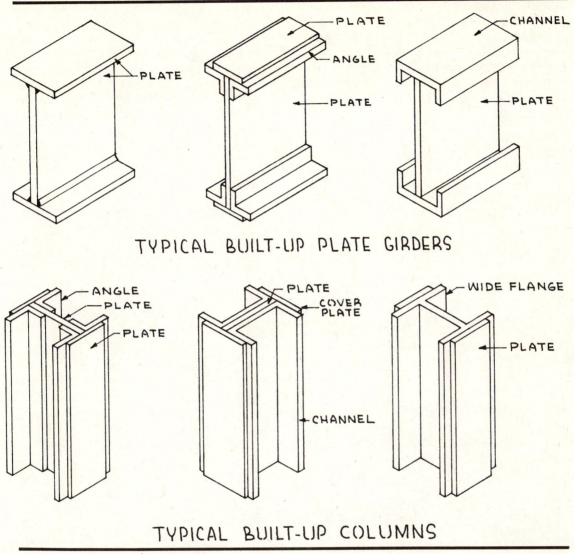

TYPICAL BUILT-UP PLATE GIRDERS

TYPICAL BUILT-UP COLUMNS

Figure 1·4 Plate girders and columns

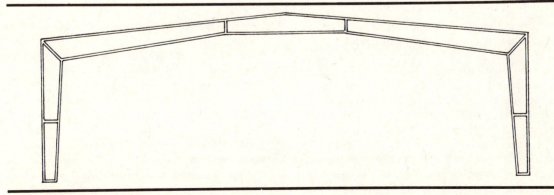

Figure 1·5 Typical rigid frame

erected, combine columns and roof beams into one rigid member. Depending upon design loads and appearance factors, such units may be based upon I-shapes, channels, or other stock components. The overall shape of the units can be greatly varied and can fit other design considerations quite well.

Some rigid frames are fabricated with varying depth (tapered), so that points receiving high stresses are reinforced. This

Trusses (Figure 1-7) are, in essence, beams with the web (vertical portion) built up with a series of short, straight components arranged to form the sides of numerous triangles rather than a single solid area. The truss is a steel framework designed to absorb (and transmit to bearing surfaces) the stresses of tension and compression.

The top member of a truss is termed the

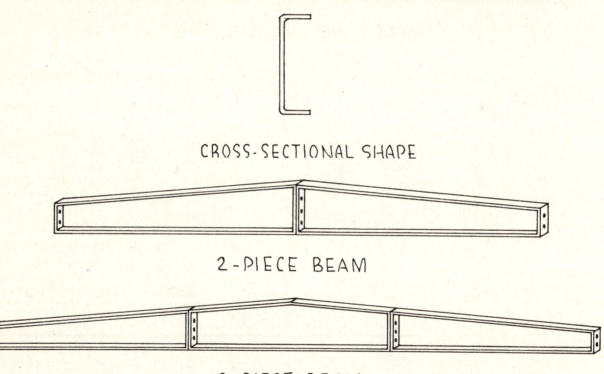

CROSS-SECTIONAL SHAPE

2-PIECE BEAM

3-PIECE BEAM

Figure 1·6 Light steel pitched beams

practice results in reduction of total weight and provides a pleasing appearance. The rapid growth in popularity of prefabricated metal buildings has led to widespread use of rigid-frame construction. These systems have also developed lightweight pitched beams (Figure 1-6) with a flat (horizontal) bottom and a sloping top which automatically builds in the roof pitch (or slope). The entire metal building concept dictates lightweight, easily assembled components which permit rapid field erection.

top chord, while the lower member is called the *bottom chord*. The two chords are held together by the triangular lacing of short members called *webs*. Web members are placed vertically and diagonally, depending upon the location in the truss and upon the loads designed for at that point. After erection, the trusses are connected horizontally and at 90 degrees to them by straight members known as *purlins*. The purlins support the roof construction and distribute the load on the trusses.

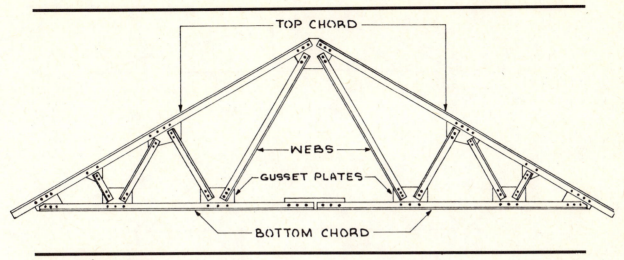

Figure 1·7   Steel roof truss

There are variations from the most common truss types, but trusses generally fall within the five main types shown in Figure 1-8: the Pratt, Warren, Fink, scissors, and bowstring configurations. The truss is most often assembled from standard sections, and a number of smaller trusses are prefabricated by some manufacturers. Trusses for long spans are designed for the specific project and are fabricated for the particular job, rather than being a stock item.

A fairly recent development, the *space frame* (Figure 1-9), offers a system for covering large areas. It is composed of straight standard sections and can be thought of as a three-dimensional truss. The standard truss utilizes the triangle as its basic geometrical unit, whereas the space frame relies upon the tetrahedron. Most space frames can be classified as triangular, rectangular, hexagonal, or isometric.

Steel domes are in frequent use and provide large areas under the roof without interior obstructions. The dome members are usually arranged in some radial fashion with additional members as parallel ribs or diagonals. Basic dome types include the systems known as *lamella* (lattice), *parallel*, *hexagonal*, *Schwedler*, and *geodesic*.

Steel cables of high-strength strands form the structural basis for many suspension structures (Figure 1-10). Suspension construction hangs floors, roofs, etc., from a central support or series of supports. Suspension construction has been somewhat limited in its application. The most extensive use has been in bridge construction.

LOADS

Every structure is subjected to loads of various types. These loads are transmitted through the structural system as specific kinds of stress and affect the individual component members. *Dead load* is the load placed on the structure by the materials of the building itself. Wind pressure on the surfaces of the structure is termed *wind load*. Areas having snow must include design calculations for *snow load*; and the weight of occupants, equipment, and furniture in the completed structure is called *live load*.

DIRECT STRESS

The simple direct stresses encountered in construction are classified by the kind of strain they produce and are designated as *tension*, *compression*, and *shear*. Individual members and component parts of structures must be designed to withstand all the types of stress to which they will be subjected.

Tension Tension is caused by two equal forces pulling on the ends of a member in

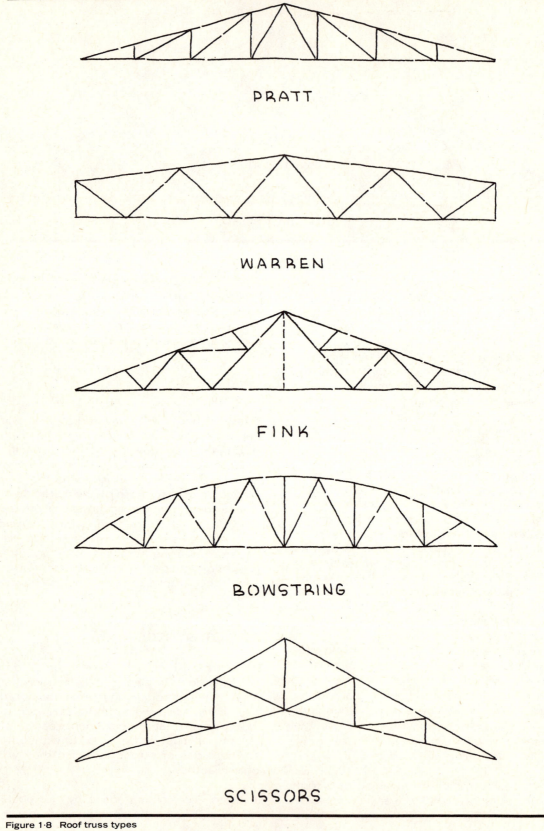

PRATT

WARREN

FINK

BOWSTRING

SCISSORS

Figure 1·8 Roof truss types

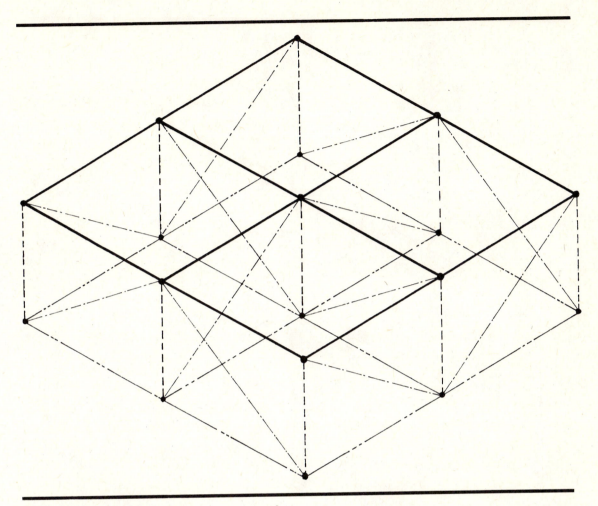

Figure 1·9   Rectangular space frame

such a way as to attempt to stretch (elongate) it. A weighted elevator cable or the anchor chain of a ship is subjected to tension, because its load tends to elongate it. A member in tension must resist a pulling action at each end (Figure 1-11).

Compression   Two opposite and equal forces pushing on the ends of a member in such a way as to attempt to shorten (or crush) it create the condition known as *compression*. Columns supporting the weight of floor or roof construction above are subjected to such shortening action. A member in a state of compression must resist a pushing action at each end (Figure 1-11).

Shear   Every time we use grass or hedge shears to trim grass or shrubs the process depends upon the condition termed *shear*.

The two blades force particles of the grass, or twig, past each other in a slipping action, with separation the result. A structural part encountering shear must resist being cut in two. The rivets or bolts joining two pieces of steel plate are subjected to the same cutting action applied on a blade of grass by the cutting edges of the grass shears (Figure 1-11).

Internal resistance to stress   The internal resistance in members to the three simple direct stresses is classified as *tensile stress*, *compressive stress*, and *shearing stress*. Beams and girders are subjected to a transverse stress and must resist bending as shown in Figure 1-11. A beam may be supported in various ways (cantilevered, at each end, etc.), but loads applied to it tend to bend it.

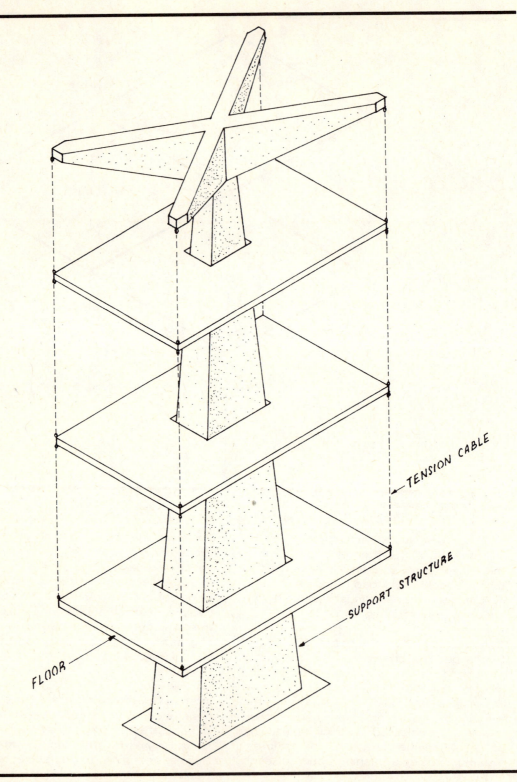

Figure 1·10 Suspension structure

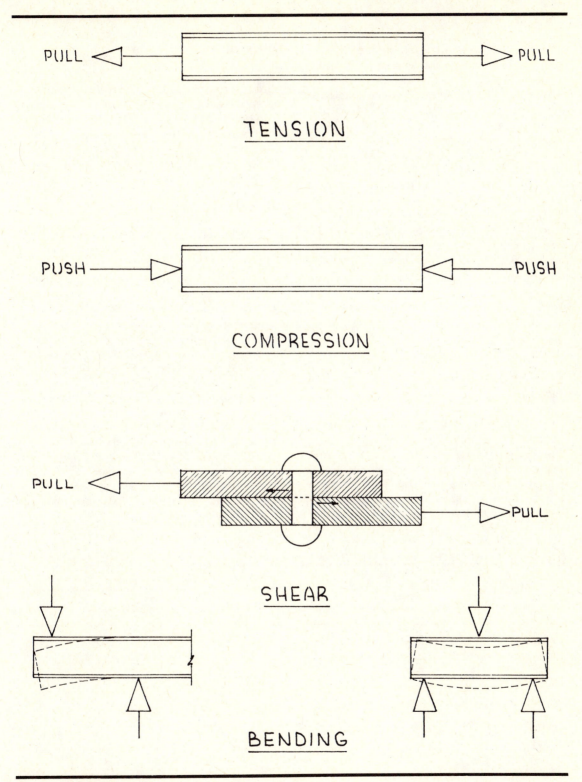

Figure 1·11 Direct stresses

UNIT STRESS

All materials vary in the amount of stress that they can withstand and in their capabilities to resist the various types of stress. To permit the designer to calculate loads, stresses, and sizes of structural members, a system of *unit stress* has been adopted. It simply spells out units of force and units of area. The direct-stress formula states that: $f = P/A$, where f is in pounds per square inch (psi), P is the load in pounds, and A is the cross-sectional area in inches. The load capability of f in psi is a unit stress. Structural designers work with a number of more complex stresses, but since this volume centers on the detailing process, their explanation is purposefully omitted.

STRENGTH

As mentioned previously, steel is strong and ductile. Its strength varies with its composition, the processes used in creating and forming it, and any special treatments to increase its stress values. The high allowable stresses of steel permit design of relatively slender members which are very much in keeping with modern architectural design. Steel is capable of withstanding higher stresses than wood or concrete, particularly in tension, and steel reinforcing bars are used to increase tensile strength of concrete.

Steel is not fireproof. It softens and loses a great deal of its strength prior to reaching the point of being red hot. To preserve the high strength potentials of steel, structural members are often covered with a thin layer of insulating concrete or wrapped with other materials such as concrete masonry or tile.

Table 1-2 shows some types of steel, their yield point, and allowable working stresses in psi. The *yield point* can be thought of in this way: when a bar is stretched, it is being elongated to more than its original length. Like a spring, if the elongation has been carried too far, the bar will not return to its original dimensions when the forces are removed. The point of loading beyond which the bar fails to return to its original size and shape is called the *elastic limit*. At a stress slightly above the elastic limit, the bar will be deformed without a proportionate increase in stress. This represents the yield point, which varies with the type of steel tested. Table 1-2 lists information outlined by the American Society for Testing and Materials (ASTM) and shows the variation in load potential for various types of steel.

TABLE 1-2 STRENGTH, YIELD POINTS, AND ALLOWABLE WORKING STRESSES FOR STEEL

ASTM Designation	Type of Steel	Thickness in Inches	Min. Yield Point, PSI	Allowable Working Stresses, PSI			
				Tension	Compression Short Lengths	Shear	Bending
A 36	Carbon steel	All	36,000	20,000	22,000	14,500	24,000
A 7 A 373	Carbon steel	All	33,000	20,000	20,000	13,000	22,000
A 440	High-strength	3/4 and less	50,000	30,000	30,000	20,000	33,000
A 441	Low-alloy	Over 3/4 to 1 1/2	46,000	27,500	27,500	18,500	30,500
A 242		Over 1 1/2 to 4	42,000	25,000	25,000	17,000	28,000

FASTENING DEVICES

Structural steel systems depend upon the adequacy of the joints and the connections of the separate components. Connections must be capable of carrying the stresses exerted upon them by the members or parts they are connecting. Essentially, fastening is accomplished by three methods: *bolting*, *riveting*, and *welding*. Each method has certain attributes which best suit it for specific types of applications. Some joints become so complex that it is difficult to provide the clearances necessary for the riveting gun or for the wrench required for bolts. Welding many components in the fabricating shop is sometimes impractical because the resulting items become too large and too heavy to be shipped with reasonable ease and economy. Many factors enter the decision to use a certain type of fastener, and many structural systems use combinations of fasteners to best meet the varying demands of the individual connections.

Riveting and welding fall within the basic category of *permanent* fasteners. Bolts are considered *removable* fasteners, because they can be removed and the assembly dismantled.

Bolt grades Bolts are available in three grades indicative of the degree of finishing they have received. *Unfinished* bolts and nuts are the roughest grade available. Manufacturing tolerances are quite large, and the surface finish is rough compared to the top grade. This grade is seldom used for structural steel work. *Semifinished* bolts and nuts are parts from the unfinished grade which have been faced (or finished) on the bearing surfaces to provide a washer face (or chamfer). These parts more closely meet the squareness tolerances demanded in high-grade products and are used for construction purposes. *Finished* bolts and nuts represent the highest commercial production standards for such parts, and are smoothly finished and dimensionally accurate. All bearing surfaces are chamfered or washer-faced.

Bolt series All three grades of bolts are available in two series, having different across-the-flats dimensions and nut thicknesses. The *regular* series contains the most commonly used sizes of bolts and nuts. The *heavy* series, used where a greater bearing surface is demanded, offers bolts and nuts with larger across-the-flats dimensions and thicker nuts.

Most common bolt Several types of structural bolts are used, but one particular bolt has become the most common of the lot. The ASTM designates it A 325, and it is a high-strength bolt. It is often used with two washers and a heavy semifinished hexagonal nut. The bolt itself is a heavy finished bolt with a hexagonal head.

Hole size The holes for all grades and series of structural bolts are generally drilled $1/16$ in. larger than the nominal bolt diameter to allow for easier positioning.

Tightness Structural bolts are tightened in place by hand wrench or torque wrench, and an often-used rule of thumb is to tighten the nut one full rotation (or turn) past the point of being finger-tight. A more sophisticated control is provided by use of a calibrated torque wrench which can be preset for desired values.

Rivets Rivets are used for structural work because they are simple, dependable, and economical. They are most widely used in shop fabrication, although field riveting is still used in some applications. Rivets are ductile steel pins, made from bar stock, with a manufactured head on one end. Once inserted in the hole, the unfinished rivet end is re-formed over the hole in the fastened part to form a second head. Rivets can be driven by methods using electric power, pneumatics, or hydraulics.

Rivet types The rivets used in construction are available with a variety of head types. Rivets available include the head types of button, high button, cone, pan,

BUTTON HEAD

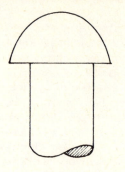

HIGH BUTTON HEAD

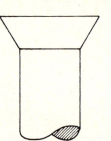

FLAT TOP COUNTERSUNK

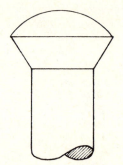

ROUND TOP COUNTERSUNK

Figure 1·12   Common rivet head types

flat-top countersunk, and round-top countersunk. Those most often chosen for structural work are the high-button head, the button head, the flat-top countersunk, and the round-top countersunk (See Figure 1-12).

Rivets in construction work fall within the large rivet class, and their shaft diameters generally range from ½ to 1¼ in. The shaft length is designed for the thicknesses of the materials to be joined, with the extra length available for forming the final head.

Riveted joint types   Two basic types of joints are riveted (Figure 1-13): lap joints and butt joints. A *lap joint* is the term applied when the plates to be fastened together are overlapped. Two kinds of lap joints are the single-riveted and the double-riveted, referring to the number of rows of rivets in the joint (one row or two rows). A

butt joint is formed when two plates are placed end to end (approximately ¼ in. apart) and fastened together with cover plates on each side. Butt joints are also designated by the number of rows of rivets used, and are called single-riveted, double-riveted, etc.

Welding processes   Welding is used frequently in joining structural components and depends upon creating a molten state of the areas to be joined, thus allowing them to fuse together. The heat required for fusion welding may come from burning gas (oxyacetylene), electric arc, or chemical reaction (Thermit). Care must be taken in welding structural parts, because the heat applied in the process may change the properties of the materials. Overheating the surrounding material can create internal stresses within the material and cause distortion of the parts being joined.

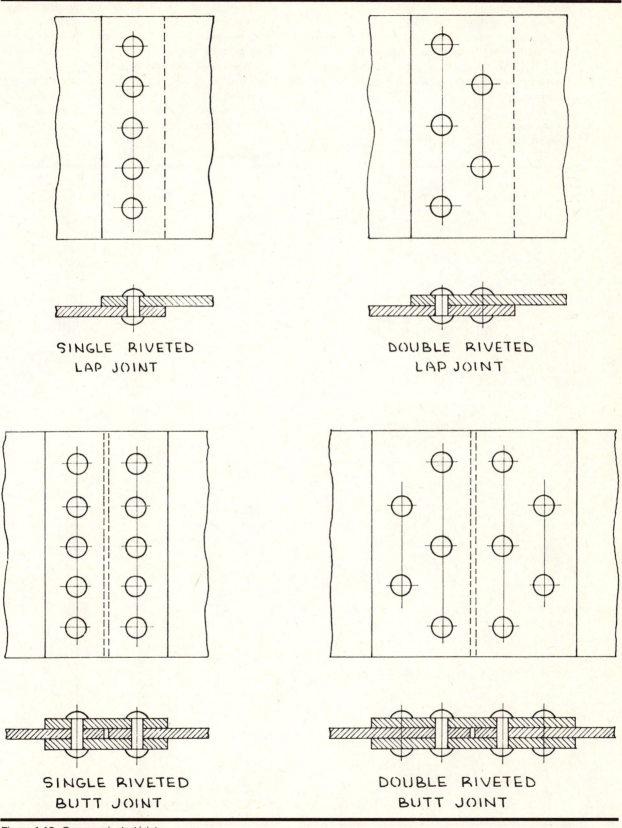

Figure 1·13 Common riveted joints

Basic welded joints and types There are five basic welding joints (Figure 1-14) commonly relied upon in structural steel work. The joints are the butt joint, lap joint, corner joint, T-joint, and edge joint. These basic joints are generally welded using one of the following basic weld types (Figure 1-15): groove weld, fillet weld, plug weld, slot weld, and spot weld. Further classifications exist for some of the basic types. For example, groove welds can be classed as square groove, V-groove, bevel groove, U-groove, and J-groove welds. Even more possibilities exist when one of the sub-types is specified on one or both sides of the material. Typical terminology in this case might be single-V and double-V groove welds. Welding symbols and terminology are discussed in more detail later in the text.

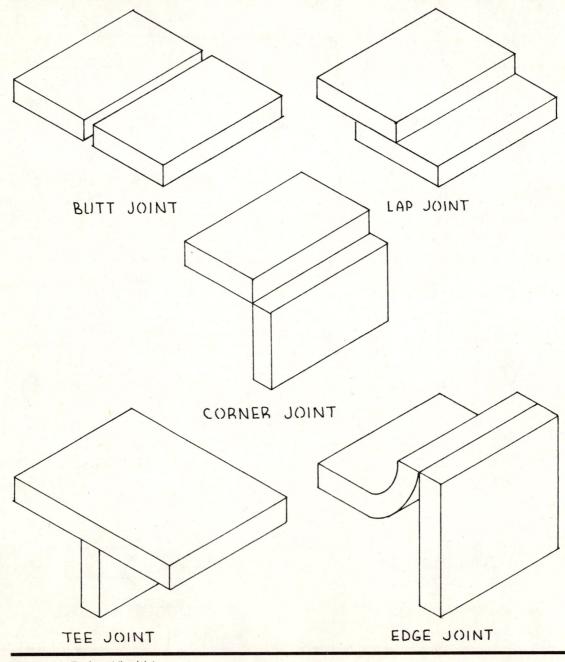

Figure 1·14 Basic welding joints

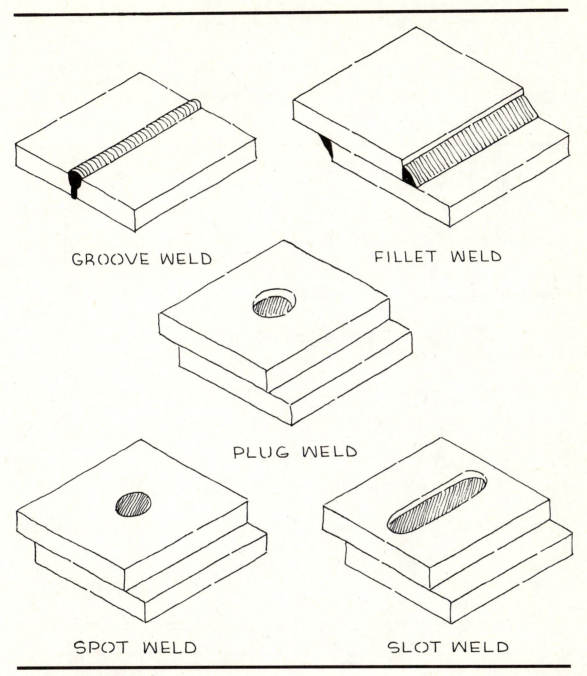

GROOVE WELD

FILLET WELD

PLUG WELD

SPOT WELD

SLOT WELD

Figure 1·15 Basic weld types

BASIC FRAMING SYSTEMS

Many methods of framing are used in building construction, and most of them conform to one of the basic types: wall-bearing, beam-and-column, or long-span framing. (See Figure 1-16.) Some projects may combine features of the basic types.

Wall-bearing construction relies upon the exterior (and interior) walls to support the end (or both ends) of such framing members as beams, joists, and rafters. The walls must be capable of carrying the floor and roof loads and are most often constructed of thick masonry. Since additional stories add weight of additional floors, and the weight of the taller wall construction, the walls at the base must be much thicker and more expensive. Use of wall-bearing construction is most often limited to low structures of one to three stories (Figure 1-17).

Beam-and-column framing uses upright members called columns to support the ends of beams and other horizontal members, thus eliminating the need for load-bearing walls. The columns are set in rows spaced to support the beams and

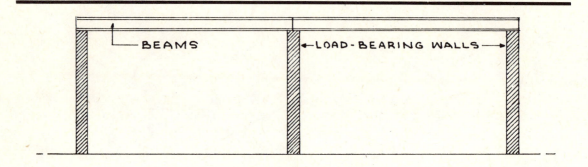

WALL - BEARING

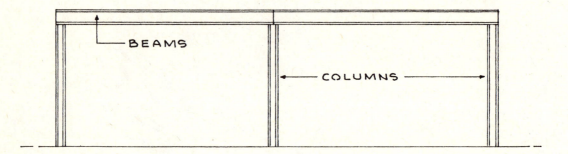

BEAM - and - COLUMN

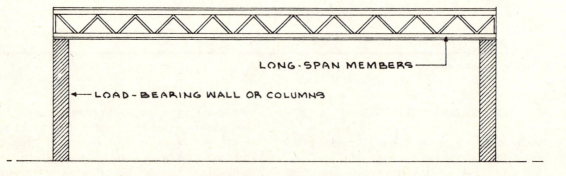

LONG - SPAN

Figure 1-16 Basic framing systems

usually form the corners of squares or rectangles in the floor plan (a grid formed by the rows). The area of space formed by the columns in each corner is termed a *bay* and is repeated throughout the system. By duplicating the bays (one on top of another), multistory buildings can be more easily constructed. Since columns transmit the building loads to the foundation, walls can be constructed in non-load-bearing form and provide for enclosure rather than support. The curtain wall of modern buildings

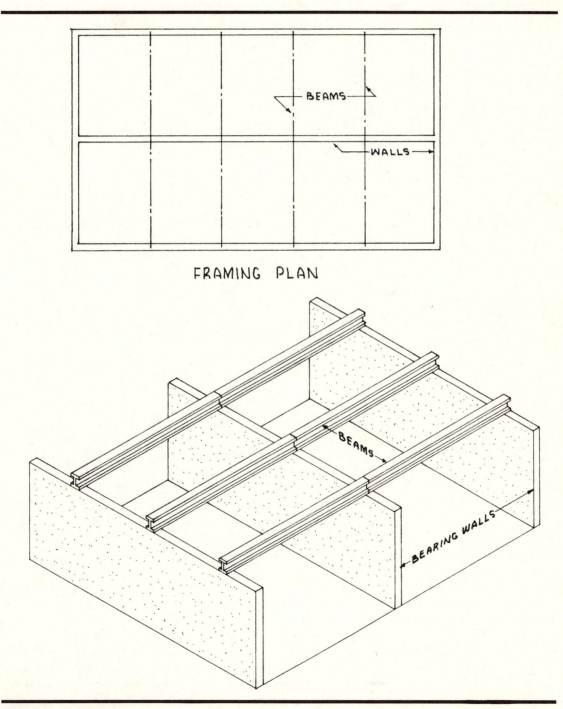

FRAMING PLAN

Figure 1·17 Wall-bearing framing

is made possible by the beam-and-column system. When constructed with wood, the same system becomes post-and-beam framing (Figure 1-18).

Long-span or clear-span framing utilizes

built-up structural members fabricated from a number of smaller structural components. Girders, trusses, and space frames permit longer spans than can be achieved with regular shapes of single structural members. These built-up members provide

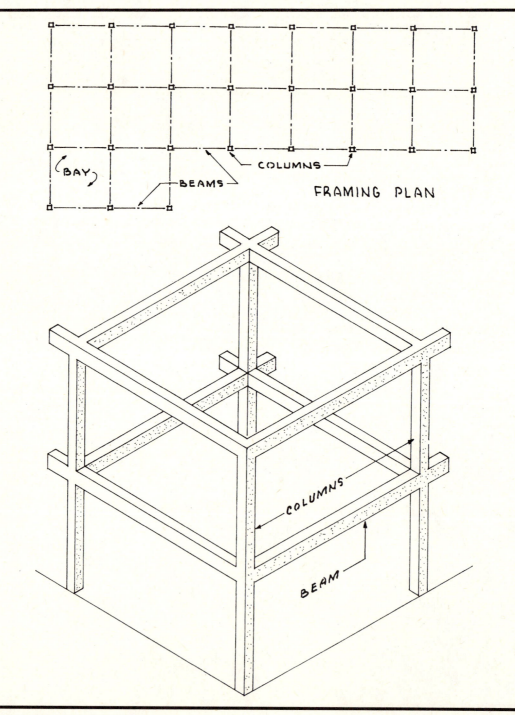

FRAMING PLAN

Figure 1·18 Beam-and-column framing

large, clear areas free of obstructing columns or walls.

Long-span members may have end-bearing on load-bearing walls or columns. The end support must be adequate to support the larger loads carried by the units, and the foundation system must provide for the transmitted concentrated loads (Figure 1-19).

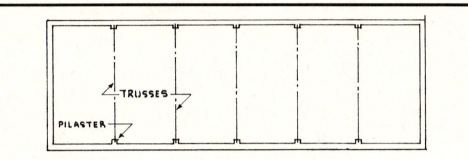

FRAMING PLAN

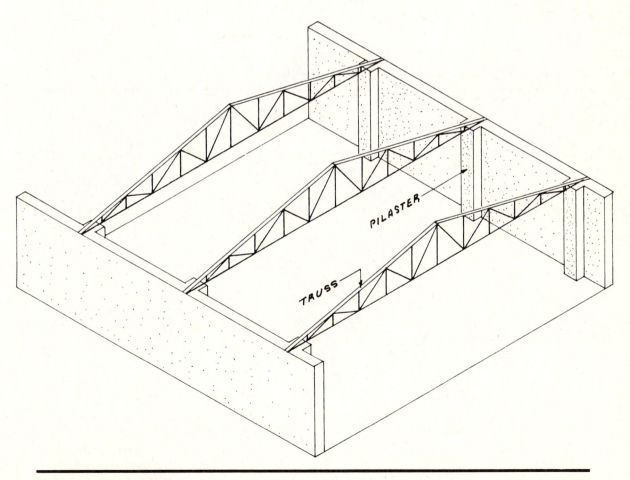

Figure 1·19 Long-span framing

SUMMARY

Steel offers excellent solutions to many of the structural design problems encountered in the construction field, but it does not negate the need for other structural materials in all instances. Exposed steel is very much subject to damage from the extreme heat generated in a major fire and fails quickly upon reaching the critical temperature range. Steel is also prone to damage by rust and corrosion, although recent developments have provided steel which is self-oxidizing and thus self-finishing.

The nature of steel makes it desirable to utilize its strength for spans of sufficient length, but the use of standard sections for extremely short spans is sometimes inefficient and cumbersome. A total steel system lacks the inherent insulating qualities of wood and concrete. In some instances exposed steel framing is inappropriate for certain types of architectural design.

Steel construction technology has successfully evolved from the first faltering efforts to use iron and cast iron as structural materials. Excellent manufacturing methods, improved quality control, and standardization of methods and sections have made structural steel dependable and easy to obtain. The construction industry has benefited from the cost savings realized through standardization and ease of erection.

Structural steel, with its rapid and efficient erection, permits framing and enclosure of buildings in very short lengths of time. Steel provides excellent long-span and multistory capabilities. Moreover, the development of space systems points to the ability of steel technology to grow in sophistication with the times.

Problems – Chapter 1

1. Sketch the standard structural steel shapes and list their names and symbols.
2. Conduct a search of manufacturers' literature and sketch the shapes of rigid frames and pitched beams available as standard production items.
3. Prepare a simple drawing of each of these trusses and label the component members: Pratt, Warren, Fink, scissors, and bowstring.
4. Prepare an end elevation of a gable-roofed structure and indicate where and how the following loads might act on the structure: wind load, snow load, dead load, and live load.
5. Sketch the truss in the illustration and indicate on each member the type of stress it is subjected to.

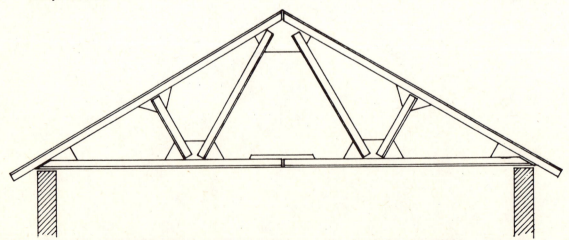

6. Examine buildings in your community and prepare freehand sketches of three of them, each of which represents one of the three basic framing systems: wall-bearing, beam-and-column, and long-span.

After Studying This Chapter You Should Be Able To:

1. Explain how steel structural systems lend themselves to contemporary residential design.
2. List the conventional steel structural systems most used in residential construction.
3. Name and explain the two main types of beam-to-beam and beam-to-column connections.
4. Discuss the importance of column base connections, and describe the methods used to provide them.
5. Describe several methods of providing end-bearing and attachment for beams in wall-bearing construction.
6. Explain how various components of prefabricated metal building systems can be used in residential construction.

14. Describe lintels and their function.
15. Describe the bill of materials, its function, and the information which it should contain.

INTRODUCTION

Structural steel is making inroads into the area of residential construction. Wood continues to dominate the residential market as a structural material, but new steel products and modern architectural design are making steel residential systems increasingly popular. Steel is frequently being used for the main structural system. It was, for many years, relegated to a subordinate role, serving solely as basement columns and girders and as reinforcing in concrete footings and foundations.

Steel lends itself well to many facets of contemporary architectural design. Wide

chapter 2
steel structural systems-residential

7. Cite the advantages of light rigid frames in modern residential structures.
8. Describe steel wall studs, their use, and their application.
9. Discuss the use of steel basement girders and columns with wood-frame residential construction, and describe the drawings necessary for their fabrication and erection.
10. Define and describe the following types of drawings: design drawings, details, and shop drawings.
11. Indicate the procedures followed in preparing design drawings, shop drawings, and details.
12. Cite the dimensioning procedures and practices used in preparing structural steel shop drawings.
13. Explain the methods of assigning marks to the various structural members on (a) design and framing drawings, (b) shop drawings, and (c) for shipping and erection.

roof overhangs are easily constructed with steel framing, and the ability of steel to span broad distances permits the use of large areas of glass. Certainly today much emphasis has been placed on modular planning and construction (Figure 2-1), and many residential designs are based on the grid. Steel beam-and-column framing facilitates the construction of spacious bays and modules, permitting the open space so much in vogue. Even the prefabricated metal building with its rigid frames is appearing (albeit disguised somewhat) on the residential scene.

This chapter explores some of the steel structural systems currently being used in residential construction, but the coverage is certainly not exhaustive. The student should realize that there can be many variations on a structural theme and should not be surprised to see systems in his own locale that are not mentioned herein. This chapter makes frequent reference to prac-

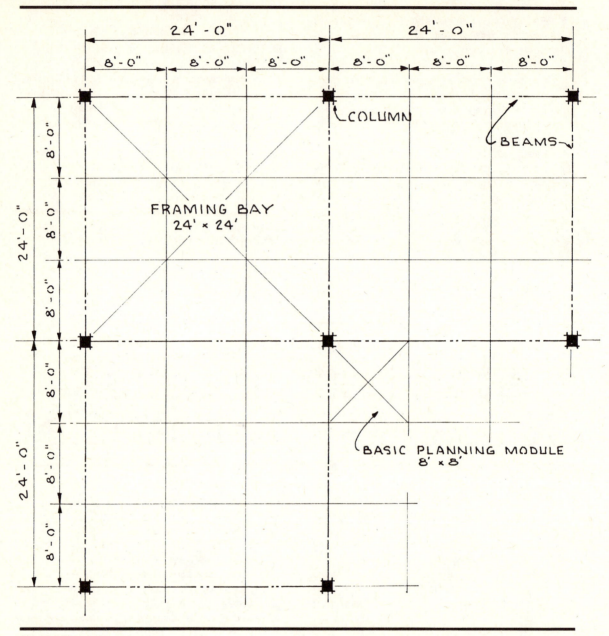

Figure 2·1 Modular plan using 8′ × 8′ grid

tices used in commercial and industrial applications, since practices in residential detailing should be consistent with those found throughout the industry.

CONVENTIONAL SYSTEMS

Several types of steel framing may appear in residential construction, including beam-and-column, wall-bearing, and long-span. Probably the most-used conventional system is the beam-and-column type. This particular system offers the residential designer a great deal of flexibility in planning and permits varying ceiling and roof heights with relative ease. The grid pattern of the bays lends itself to modular planning and construction, and the relatively wide column spacing facilitates open planning and freeflowing interior space (Figure 2-2).

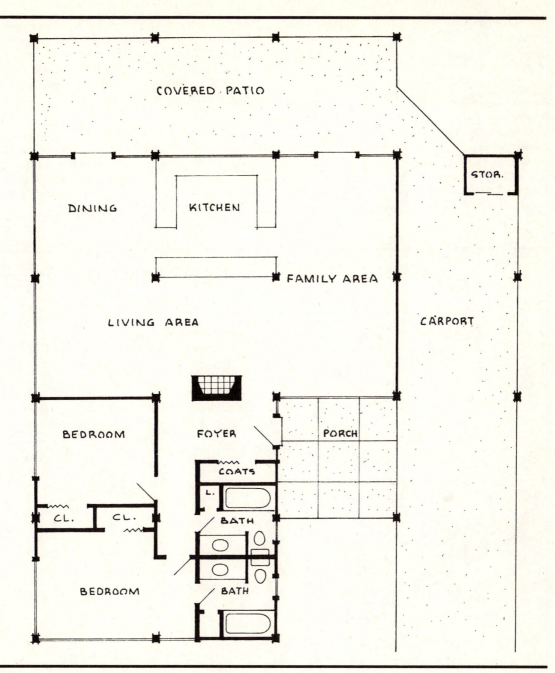

Figure 2·2 Beam-and-column residence

BEAM-AND-COLUMN FRAMING

As the name implies, beam-and-column systems combine vertical members (columns) with horizontal members (beams) to carry roof and upper floor loads. The columns transmit loads directly to the ground and are placed upon footings (concrete pads) to allow the load to be sufficiently distributed over a larger area (so that the soil can support the structure). Beams (usually running in two directions) are fastened to the columns. Spanning between beams, and providing the actual construction system for the floor or roof themselves, are smaller horizontal members called *purlins* (or joists). One variation used in steel construction is that of placing ribbed steel decking directly over the beams to form purlins and deck in one unit. (See Figure 2-3.)

Beam connections The method of joining beams to beams and beams to columns has two general configurations, called the *framed connection* and the *seated connection* (see Figure 2-4). The framed connection relies totally upon two short angles attached to the web of the beam and to the other beam or column. The seated connec-

tion uses the angles of the frame connection and provides an additional angle under the end of the beam. The bottom angle, attached to the column or other member, provides a shelf upon which the end of the beam rests.

The seated connection is advantageous in that it provides support for members while they are being aligned and fastened during field erection. It also makes it possible to reduce the number of rivets or bolts required in the fastening angles.

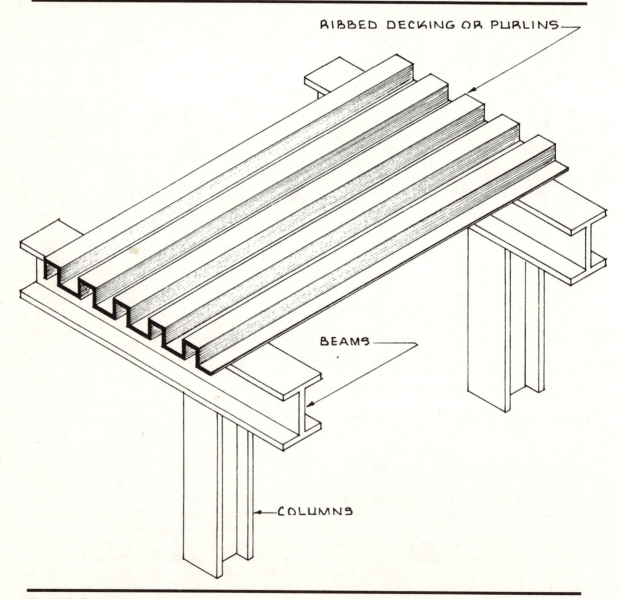

Figure 2·3 Beam-and-column framing

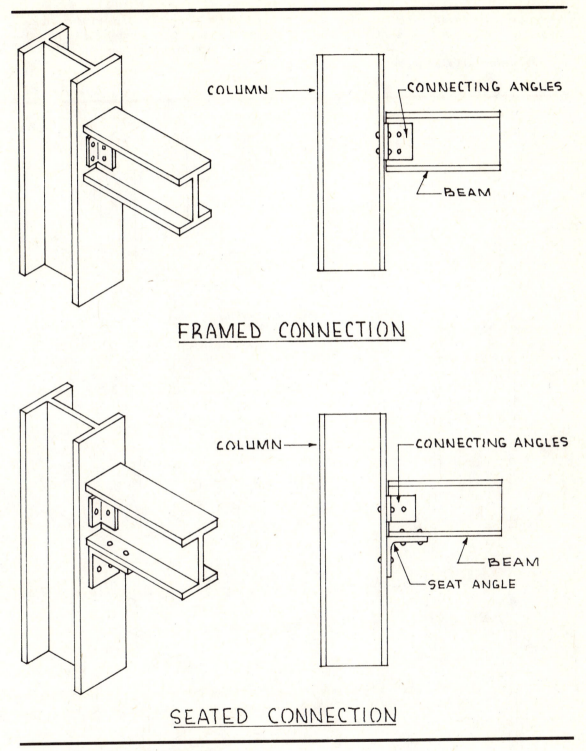

Figure 2·4 Framed and seated connection

Framed beam connections have been stand-ardized, and tables are available which provide information for sizes of the angles, number of sizes of rivets or bolts required, and the allowable loads when used with specific structural sections.

Welding is sometimes used to attach the angles of either frame or seated connections to the beam. Field welding is not always preferred to bolting, and even when connecting angles are welded to beam ends, the angles are often bolted to the column or adjoining member.

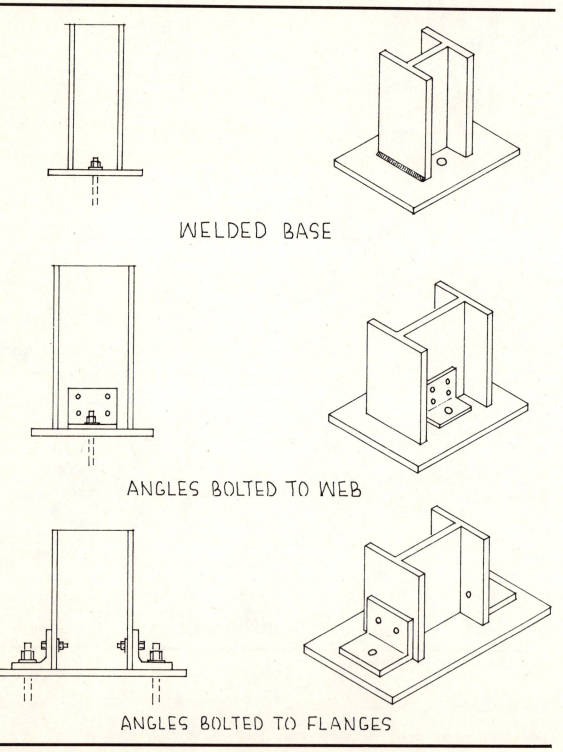

WELDED BASE

ANGLES BOLTED TO WEB

ANGLES BOLTED TO FLANGES

Figure 2·5 Column base plates

Column base connections Connections of column bases are varied, depending upon the particular type of application and the types of stress to be encountered. Several types of column base plates are shown in Figure 2-5. Most of these systems rely upon a heavy steel plate (either welded to the column or with brackets welded to the plate for bolting to the column in the field), which will be bolted to the footing or foundation with heavy steel anchor bolts. Generally the longer dimension of the base plate runs parallel to the web of the column.

It is difficult to place plates on concrete as it is being poured and provide the necessary degree of accuracy. Rough concrete is usually held 1 in. below the desired elevation of the bottom of the plate. Plates are placed, leveled, and supported by shims, with grout (high-strength concrete) carefully grouted under the plate to provide an accurate contact area. In instances where very large plates are to be grouted, holes are provided in the base plate so that grout can be poured into the central area under the plate.

Anchor bolt holes and plates The holes in the plate for the anchor bolts themselves are enlarged $\frac{5}{16}$ to $\frac{9}{16}$ in. more than the bolt diameter to provide necessary adjustment during setting. The base plate enlarges the area transmitting the loads from column to foundation so that the concrete will not be crushed by the highly concentrated load. It is desirable to have base plates which are perfectly flat. The designer may call for the plates to be straightened in a press or, in the case of heavy commercial construction, machined for an accurate surface. Commercial specifications may call for the ends of the columns to be milled as well.

WALL-BEARING STRUCTURAL SYSTEMS

Residential construction sometimes depends on wall-bearing systems incorporating heavy-load-bearing masonry walls and standard steel beams or bar joists. A number of designs have been created with masonry walls as a prominent feature of the decor (see Figure 2-6), and the resulting system is simple and direct. The

ends of the structural members can be easily cantilevered (extended) past the exterior wall line to create large roof overhangs. Particular attention must be given, when detailing a wall-bearing steel system, to the details of connections and end-bearing conditions for the steel members.

End connections The details in Figure 2-7 represent typical examples of the details for fastening the ends of beams and bar joists. Beams are sometimes provided with attached anchoring devices which are simply embedded in the masonry work as it is laid. Other systems rely upon connecting angles which can be attached with anchor bolts at the wall. The beams are then placed and bolted to the angle connectors.

LONG-SPAN FRAMING SYSTEMS

Steel long-span systems are limited in their application to residential construction. The relatively small spans common to residential design make the larger steel trusses, girders, and long-span joists uneconomical for use in such construction.

Rigid-frames Light commercial prefabricated metal buildings are finding increased use in the residential market. The structural members of these rather light systems more closely approximate the dimensions encountered in residential work and provide unobstructed interior space. Erection of rigid frames does not require specialized or heavy equipment, and the building erection and enclosure can be quickly accomplished.

Part of the impetus for use of prefabricated metal buildings for residential work is due to the new types, textures, and colors available in prefinished wall and roof panels. In many instances, however, the steel structural system alone is utilized in conjunction with conventional materials for wall and roof construction.

STEEL WALL STUDS

The prefabricated steel studs (Figure 2-8) found in commercial structures are some-

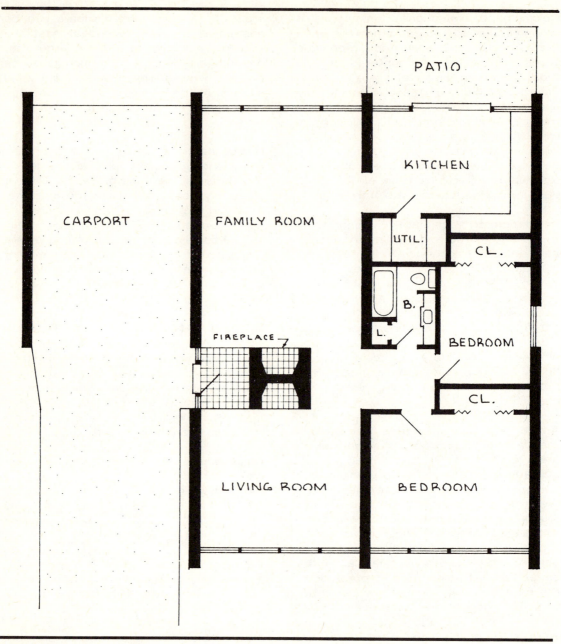

Figure 2·6 Wall-bearing residence

times used in residences for interior wall framing. These units, when faced with incombustible finished wall materials of sufficient thickness, provide adequate fire-resistance ratings. A usual installation consists of the steel studs covered with metal lath on each side and plastered with ¾ in. of plaster over the metal lath. Clips and other fastening devices are available for attaching gypsum lath (rock lath), gypsum wallboard (Sheetrock), or paneling directly to the studs. Steel studs are particularly suitable for creating tall walls, and with proper stud sizes and spacing, and horizontal stiffener channels, walls as high as 26 ft can be easily erected.

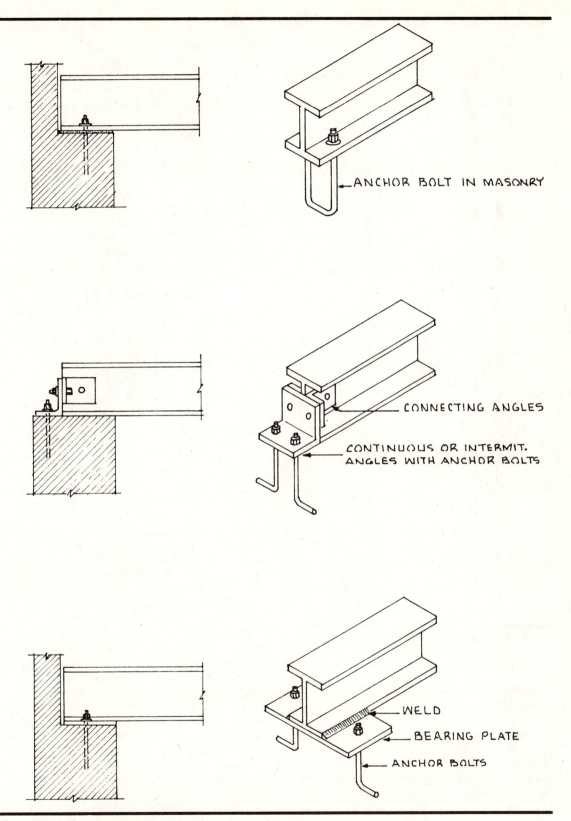

Figure 2·7 Typical end connections

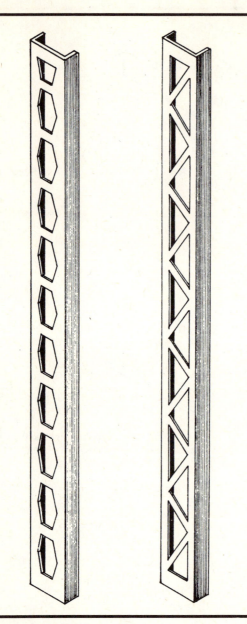

Figure 2·8 Typical steel wall studs

BASEMENT COLUMNS AND GIRDERS

Basement columns and girders reduce the span required for floor joists and provide end-bearing for joists at stairwells and other openings (Figure 2-9). This is the most frequent use of steel components in residential construction, along with steel reinforcing for concrete walls and footings, and lintels over window and door openings.

Columns and girders The girder generally runs parallel to the long dimension of the house and usually falls near the center line for the space. The columns are spaced according to the loads and spanning capability of the section selected for the girder. Many residences rely upon a wood beam (or girder) placed on top of steel columns. Columns for wood beams must be spaced closer together because of the span limitations of wood. Column spacings for wood girders built up from two or three units of dimension lumber (bolted 2 × 6s, 2 × 12s, etc.) range, in general, from 5 ft 6 in. to 12 ft 6 in. center to center. Steel girders span much larger distances depending upon the section specified.

End-bearing Basement girders are provided with end-bearing by two methods. (See Figure 2-10.) One method calls for a projection of the foundation wall (called a pilaster) to provide a surface for the girder end to rest on. The other, simpler method requires a notch (recess) formed in the wall itself to receive the end of the girder.

Basement columns Steel columns for supporting basement girders usually have a bearing (base and cap) plate tack-welded to each end. The type of fastening system depends upon the particular application. The most recommended practice provides pad footings below the level of the basement floor slab. The column base plate is set on a grout (or mortar) bed on the footing, with the floor slab poured over the plate and around the pipe column. The desirable minimum distance from top of plate to top of slab is 2 in.

Column cap (or top) plates attach to the steel girder with bolts or to wood girders with bolts, lag screws, or spikes. Channel sections are sometimes used as cap plates to provide a stirrup for the girder, in which case attachment to the girder can be from below or through the sides. (See Figure 2-11.)

The columns usually specified for residential application are hollow round pipes of standard structural steel. The designer may elect to have the pipes filled with concrete, and they thus are Lally columns. Screwjack-equipped pipe columns are

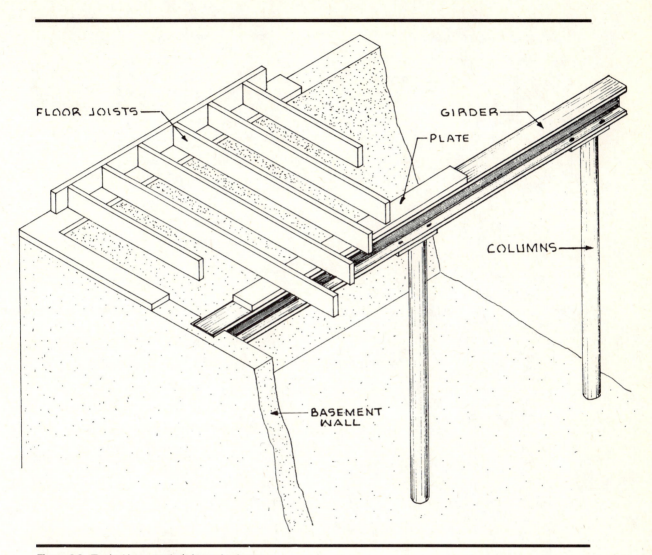

Figure 2-9   Typical basement girder and columns

available (sometimes termed floorjack or adjustable post) which permit adjustment of the girder elevation during and after construction. The fixed standard column is used with much more frequency.

DESIGN AND DETAIL DRAWING

In drawing the basement girder and column system, the items are shown on the basement (or footing and foundation) plan or on the floor-framing plan. The typical basement plan in Figure 2-12 shows the girder with a heavy broken line (center line). The columns are indicated as circles and called out by note. The required distances are di-

mensioned to the center line of the girder and to the center lines of the pipe columns. The girder end conditions (shown here for both types of conditions) are usually identical and are drawn in plan view with information called out by note. The elevation of the top of the steel beam or girder is called out on the plan.

Details of attachment of the items are prepared at a larger scale (1 in., 1½ in., 3 in. = 1 ft 0 in., etc.) and show joints, connections, and splices (where applicable). Such details are most often drawn as a cross-sectional view, chosen to best explain the intended method of assembly.

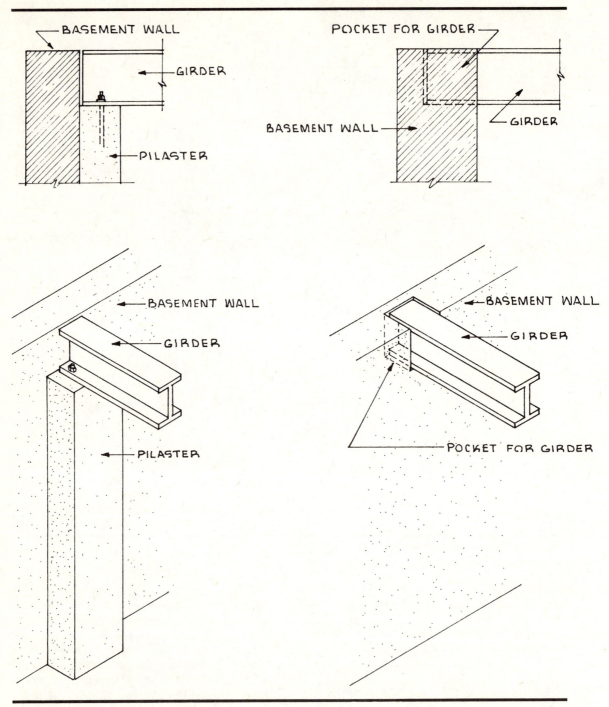

Figure 2·10 End-bearing for basement girders

The actual production of detailed drawings for the steel items used in residential construction follows the procedures used in detailing complete steel structural systems. The design drawing (in the example of Figure 2-12 combined with the basement plan) calls out size, spacing, and pertinent information about the members. The details show the procedure for assembly, and the shop drawings present details of how the various parts are to be fabricated for assembly.

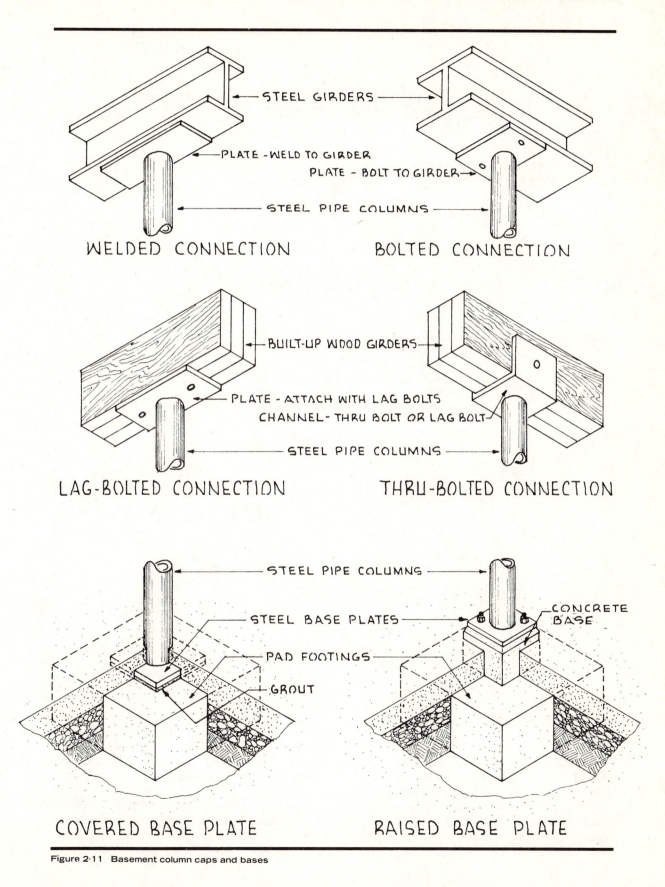

WELDED CONNECTION

BOLTED CONNECTION

LAG-BOLTED CONNECTION

THRU-BOLTED CONNECTION

COVERED BASE PLATE

RAISED BASE PLATE

Figure 2·11 Basement column caps and bases

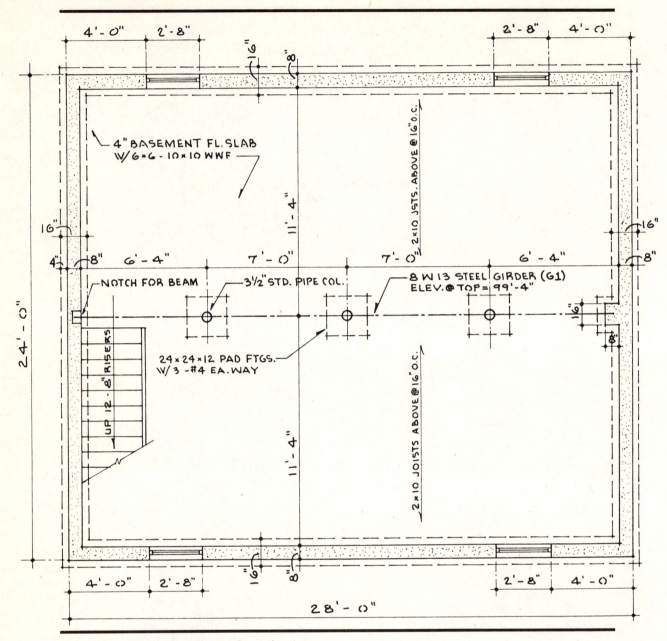

4'-0"   2'-8"   2'-8"   4'-0"

4" BASEMENT FL. SLAB
W/ 6×6 - 10×10 WWF

16"

11'-4"

2×10 JSTS. ABOVE @ 16" O.C.

16"

4" 8"   6'-4"   7'-0"   7'-0"   6'-4"   8"

24'-0"

NOTCH FOR BEAM   3½" STD. PIPE COL.   8 W 13 STEEL GIRDER (G1)
                                                ELEV.@ TOP = 99'-4"

UP 12 - 8" RISERS

24×24×12 PAD FTGS.
W/ 3 -#4 EA. WAY

11'-4"

2×10 JOISTS ABOVE @ 16" O.C.

4'-0"   2'-8"   2'-8"   4'-0"

28'-0"

Figure 2·12   Basement or footing and foundation plan

SHOP DRAWINGS

Figure 2-13 shows how the girder called out on the basement plan might be drawn in the shop drawing. Notice that the hole conditions at the end and attachment points over the columns have been detailed. The girder itself is shown in front elevation and is drawn to a scale of 1 in. = 1 ft 0 in. except for the length, which is foreshortened.

In structural drawing no break line is used to indicate the foreshortening. By the same token, small distances are sometimes exaggerated in order to present a more accurate picture of the details.

The shop drawings for the columns shown in the basement plan are indicated in Figure 2-14. These drawings detail the columns, their base plates, and their cap

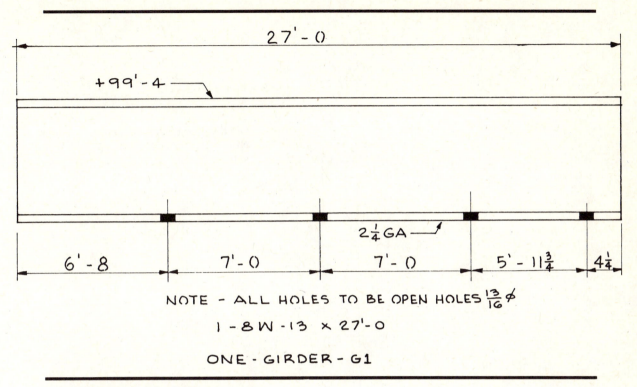

NOTE – ALL HOLES TO BE OPEN HOLES $\frac{13}{16}$ ∅

1 – 8W – 13 × 27'– 0

ONE – GIRDER – G1

Figure 2·13 Shop drawing for basement girder

plates. Columns can be drawn either vertically or horizontally on the sheet. If the column is drawn upright, the base is at the bottom of the drawing. If the column is drawn horizontally, the base is on the left side of the sheet.

BILL OF MATERIALS

The bill of materials is an important part of most sets of shop drawings. This item assists the shop personnel in locating, cutting, and assembling all of the component parts and materials necessary for the shipment to the job site. Figure 2-15 shows a typical bill of materials. Notice that the items covered include the assembling mark, number of pieces, shape, section, length, weight per foot, total estimated weight, and space for special remarks. The bill of materials is often combined with a shipping list or a mill order. A shipping list contains spaces to enter number of pieces, description mark (beam, column, etc.), actual weight, and shipping record. Mill order forms include items which are not in stock and which must be ordered directly

from the steel mill. These forms (Figure 2-15) contain spaces for the number of pieces, shape and section, length, and item.

Bills of materials must be accurate, since they form the basis for the operations of the persons charged with collecting, fabricating, and shipping the materials; they also assist in the determination of the prices charged for the job by the steel supplier.

The student may have a feeling of bewilderment when faced by the overwhelming bulk of information required for the structural detailing process, but there are very helpful aids in the manuals available for use in the process. Complete coverage of shapes, sizes, weights, actual dimensions, joints, connections, and fasteners in these manuals provides a tremendous assist.

DIMENSIONING

Dimensioning procedures in structural shop drawing vary from some practices

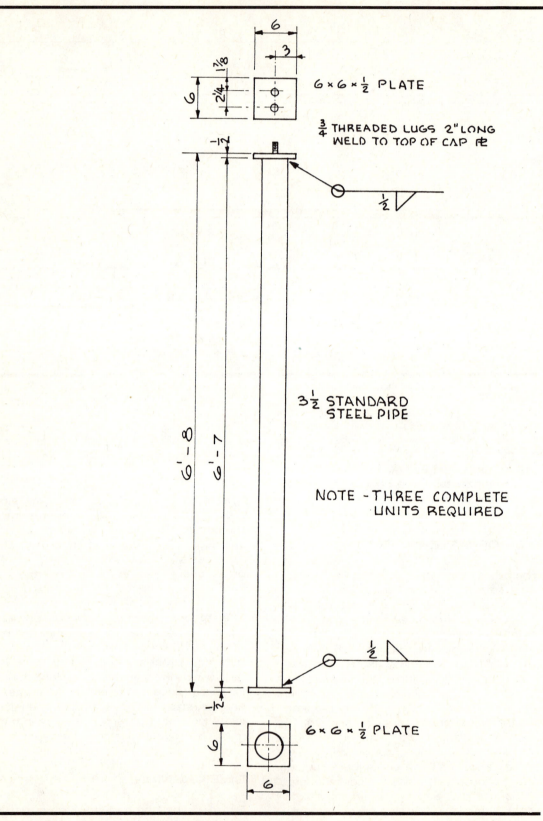

6 × 6 × ½ PLATE

¾ THREADED LUGS 2" LONG
WELD TO TOP OF CAP ℞

3½ STANDARD
STEEL PIPE

NOTE - THREE COMPLETE
UNITS REQUIRED

6 × 6 × ½ PLATE

Figure 2·14 Shop drawing for basement column

BILL OF MATERIALS

LINE	NO.	SHAPE				LENGTH FT.	LENGTH IN.	ASS'BLY MARK	REMARKS	WEIGHT		NO.	SHAPE	LENGTH FT.	LENGTH IN.	ITEM
		SHOP BILL										MILL ORDER				
1	2	14	W	84		27	4	a								
2	1	14	W	43		14	8	b								
3	1	14	W	61		14	8	c								
4	4	℡	8	½		1	4	d								
5	2	L	6	4	¾		10									
6	2	L	6	6	¾		11½									
7	1	℡	12	¾												
8	1	8	S	23												
9	2	12	W													
10																
11																
12																
13																
14																
15																
16																
17																
18																
19																
20																
21																
22																
23																
24																

Figure 2·15 Partial example of bill of materials

observed in machine drawing. The following basic dimensioning guidelines most often apply in shop drawing (see Figure 2-16):

1. Dimensions should be placed far enough away from the view of the object to avoid crowding the drawing itself.
2. Dimensions should build from the object drawing, with smallest dimensions closest and longest, and overall dimensions farthest away.
3. Dimensions should usually be referenced from the center lines of beam shapes, the backs of angle shapes, the backs of channel shapes, and the center lines of round shapes.
4. Dimensions should be given to the point on the beam which is designated

by level (or elevation). In most cases this means dimensioning to the top or bottom of the beam, rather than to top and bottom both.

5. Dimension lines on structural drawings run unbroken for their full length,

with dimensions placed above the dimension line.

6. Dimension figures generally carry the mark for feet, but the inch mark does not appear: thus 8'-8 stands for 8 ft 8 in.

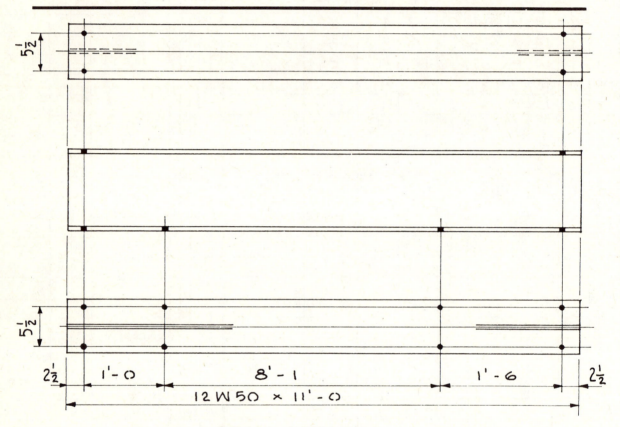

BEAM DETAIL

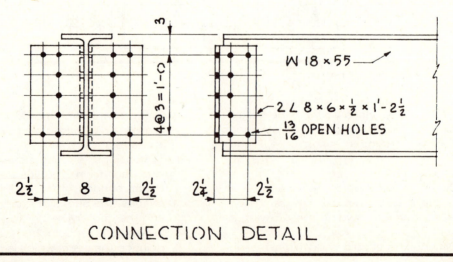

CONNECTION DETAIL

Figure 2·16 Examples of Dimensioning

7. Unlike many other systems, structural dimensioning for units less than 1 ft uses only the number to designate inches. The inch is assumed, so that 0'−5 becomes simply 5.

8. The reverse is true for dimensioning an even number of feet where no inches occur; thus 8 ft even appears as 8'−0.

9. As with any dimensioning system, dimensions should be provided for all items of significance, and crossed dimension lines should be avoided wherever possible.

10. Dimensioning systems for holes are generally of two types. One system utilizes extension dimensions starting from the center line of hole groups. The other system gives extension dimensions to the first line of holes in each group, progressing from left to right.

IDENTIFYING PARTS

All components of the structural system should be clearly identified, both on framing and design drawings and on shop drawings. There is no totally uniform system in general use for assigning marks to the various parts. The system which perhaps is most used assigns a capital letter indicating the function of the piece, followed by a number which designates its location in the framing plan. Letters assigned are given as B for beam, C for column, G for girder, and L for lintel.

The small angle connectors are usually assigned small letters, and all similar and identical items on any single drawing will carry the same letter designation. To avoid the possibility of confusion, only a, b, c, d, f, h, k, m, n, p, s, t, v, and w are normally used for this purpose. When a single sheet requires more than the fourteen letters available for marking, the letters are doubled for the next items; thus additional parts might be marked aa, bb, etc.

Shipping and erection marks are necessary to avoid confusion during the erection process at the building site. The aforementioned marking systems provide locational data and are marked on the actual struc-

tural member. In multistory construction, beam marks often carry the floor level (first, second, third, etc.) for which the beam is intended. A number 10 beam used for both a third- and fourth-floor application would have the floor number (framing level or tier) following it, encircled as 3 , 4 , etc. (See Figure 2-17.)

Figure 2-17 also shows how columns might be marked. Notice that one way of marking a single-piece column which extends for more than one floor is to designate the column and add the number of tiers in parentheses, with 0 representing the base plate (lowest) level and 1, 2, 3, etc., representing subsequent floor levels or tiers. As discussed previously, identical beams may be given the same number, but every column (even though some may be identical) is given its own unique marking symbol.

DETAILING LINTELS

Steel lintels in residential work serve to support materials above such openings as doors, windows, and archways. The most common application is to carry the brick veneer over such openings, and the steel angle section lends itself particularly well to this use.

The lintel is represented on the residential floor plan in a similar manner to the representation used for steel girders and beams (see Figure 2-18). The sizes of the sections may be called out by note at the location or by a very brief lintel schedule. The details of the lintel are not usually drawn separately in residential plans but are shown in the sectional details drawn for the head conditions of doors, windows, and openings.

DETAILING MISCELLANEOUS STEEL MEMBERS

Miscellaneous beams and columns are used in frame construction for providing clear space in garages and extremely large rooms. These items are handled in the same way as the girders and columns in residential basements.

The most natural and common use of structural steel in frame construction is to provide more adequate support of loads and construction over large distances and above wall openings. The use of built-up wood members for that purpose is often time-consuming, expensive, and results in a bulky appearance.

EXAMPLES OF WALL-BEARING SYSTEMS

Wall-bearing construction in residential work sometimes combines standard steel beams with heavy masonry walls (see Figure 2-19). The walls may be of stone, brick, or concrete block, and are often a combina-

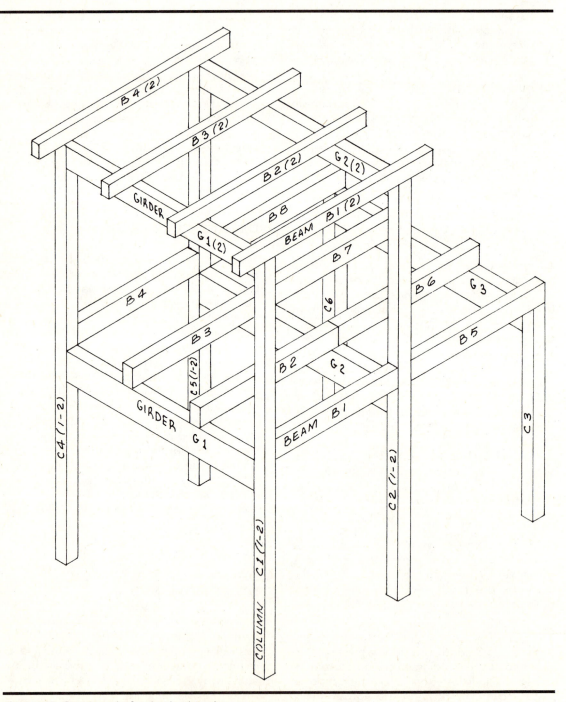

Figure 2·17 Sample marks for structural members

tion of two of these materials. In such a system, the walls serve as enclosure and as the basis for supporting roof construction and loads, providing the end-bearing for upper floor and/or roof beams.

Residential systems   Figure 2-19 shows how some types of contemporary resi-

dences are designed based upon the wall-bearing system. The use of such a system has gained in popularity, particularly in the multifamily dwelling unit (row housing) and in the motel construction field. It provides a fairly easy way of obtaining a good degree of permanence and a satisfactory fire-resistance rating. The exposed masonry

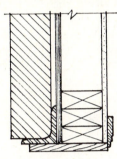

SINGLE ANGLE - FRAME CONST.

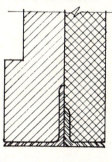

UNEQUAL ANGLES BACK-TO-BACK

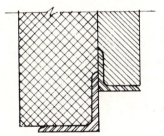

UNEQUAL ANGLES OFFSET

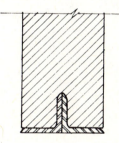

EQUAL ANGLES BACK-TO-BACK

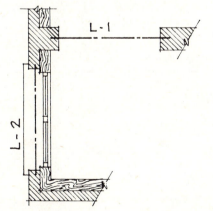

REPRESENTATION IN PLAN

STANDARD BEAM WITH PLATE

Figure 2·18   Examples of lintels in plan and section

does not require refinishing, and the walls provide excellent sound barriers between sections of the residence or between adjacent dwelling units. Certainly some limiting factors are the problem of providing large openings in the load-bearing walls and the later difficulty encountered in extensive remodeling which might require additional openings or relocation of major walls.

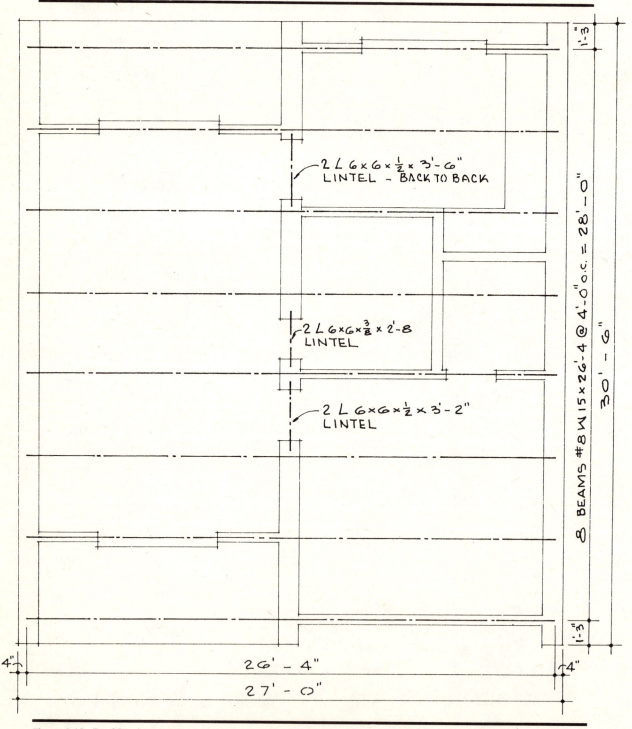

2 L 6×6×½ × 3'-6"
LINTEL - BACK TO BACK

2 L 6×6×⅜ × 2'-8
LINTEL

2 L 6×6×½ × 3'-2"
LINTEL

8 BEAMS #8 W15×26-4 @ 4'-0" o.c. = 28'-0"

3'-1"

30'-0"

3'-1"

4"

26'-4"

27'-0"

4"

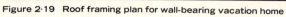

Figure 2·19   Roof framing plan for wall-bearing vacation home

Foundations and walls Particular care must be taken in planning such structures to provide a sufficiently large foundation system to accommodate the heavy wall loads. The walls must be adequate to with- stand settlement, wind deflection, and the tendency to tip. The points where the beams rest (or bear) on the walls must be designed to resist movement and the tendency to crush the wall material with the

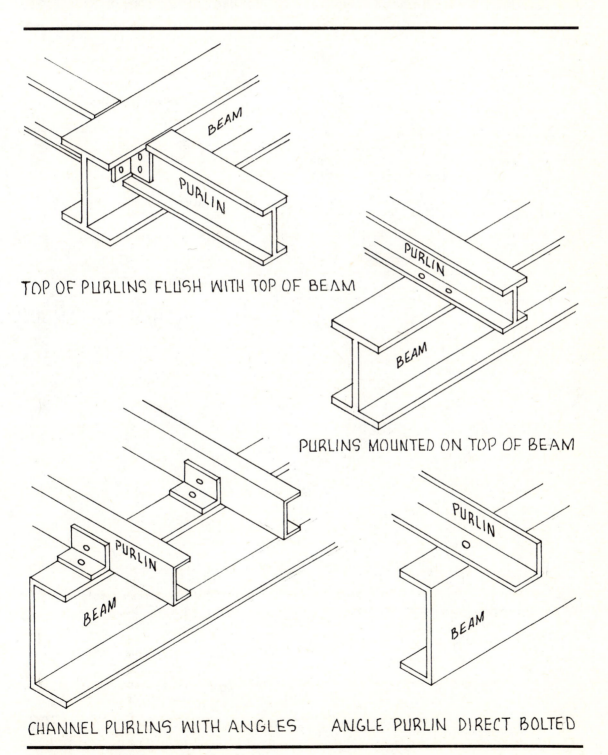

TOP OF PURLINS FLUSH WITH TOP OF BEAM

PURLINS MOUNTED ON TOP OF BEAM

CHANNEL PURLINS WITH ANGLES

ANGLE PURLIN DIRECT BOLTED

Figure 2·20 Examples of purlins and connections

heavy roof and floor loads transmitted by the beams.

Roof beams The roof beams themselves must be of sufficient size to resist deflection in bending (sagging) them downward in the center. Since such forces are exerted upon the beams, it is natural that the end connections must be capable of resisting the force of the beam ends tending to kick upward. The roof system constructed over the roof beams may be one of several con-

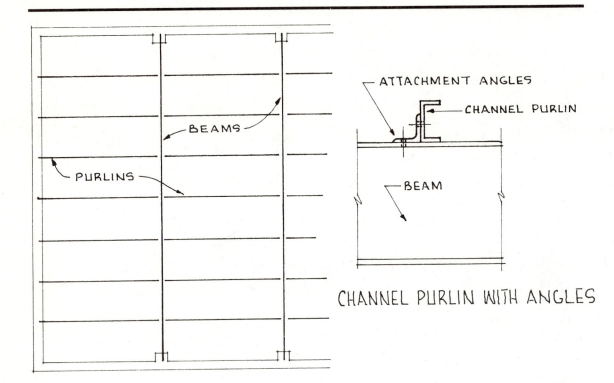

REPRESENTATION ON FRAMING PLAN

CHANNEL PURLIN WITH ANGLES

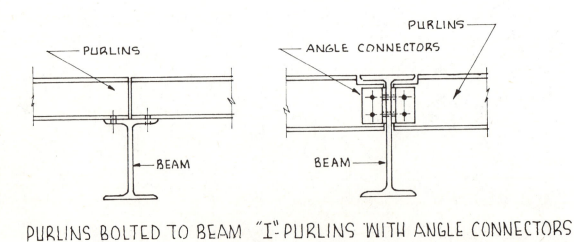

PURLINS BOLTED TO BEAM   "I" PURLINS WITH ANGLE CONNECTORS

Figure 2·21 Representation of purlins in plan and details

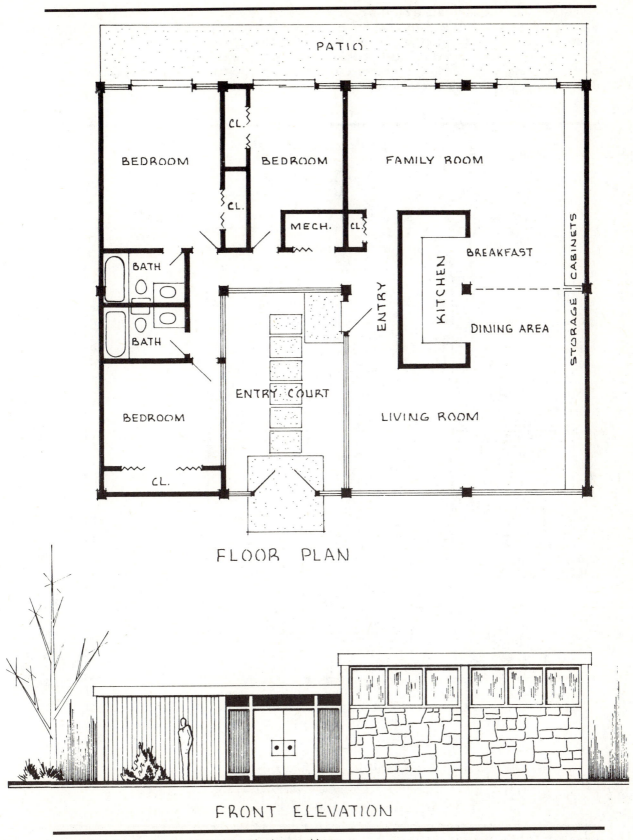

FLOOR PLAN

FRONT ELEVATION

Figure 2·22 Display drawings for beam-and-column residence

figurations. Many systems use purlins over the main beams, which in turn support a deck of steel, wood, or other material. Some systems place a ribbed deck (combining decking with purlins) of steel or prestressed concrete directly over the beams. Even heavy laminated or solid tongue-and-groove wood decking (capable of spanning 16 ft in the thicker sizes) has been used with steel beams, eliminating the need for purlins and providing a finished ceiling at the same time.

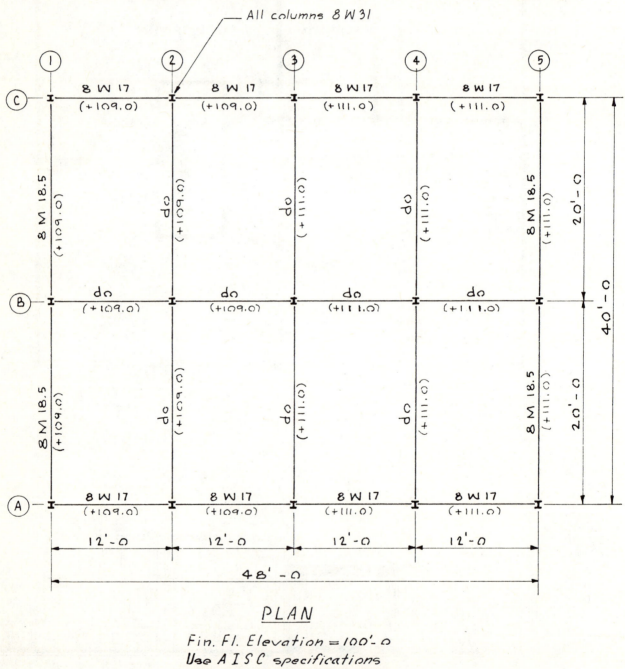

Figure 2·23 Framing plan (design drawing) for beam-and-column residence

Purlins and decks When the system utilizes purlins (or some type of ribbed deck acting as purlins), these components are actually mini-beams providing support for the roof between the main beams. They may be of standard beam shape in a

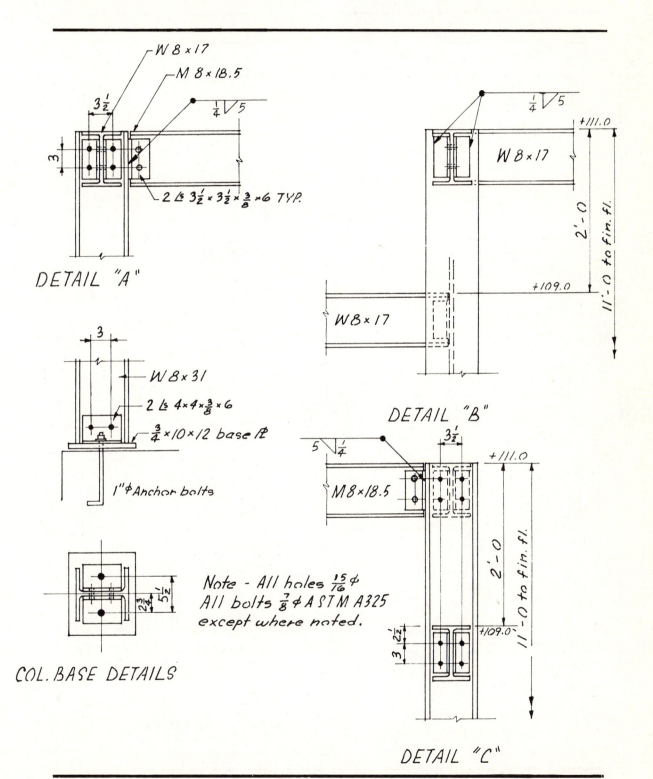

Figure 2·24 Typical details for beam-and-column residence

smaller, lighter section, or, in some cases, the light channel, angle, or Z-sections can serve this purpose quite well.

Figure 2-20 provides a view of several systems for spanning the distance between beams and supporting the actual

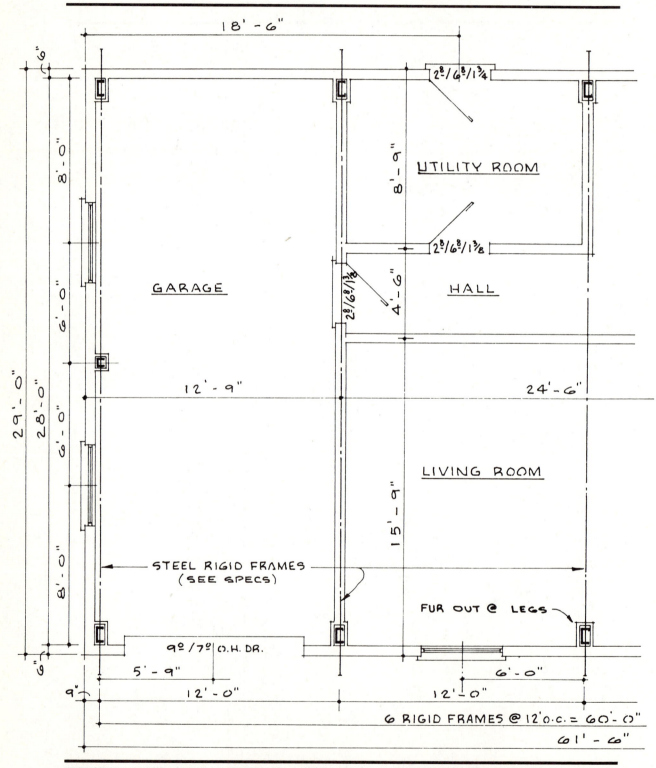

Figure 2-25 Partial floor plan for residence with steel rigid frames

roof deck. Figure 2-21 indicates the method of detailing the various members. It should be quite obvious that in residential load-bearing construction many (if not most) members are identical (or similar and reversed), thus greatly simplifying the detailing process.

EXAMPLES OF BEAM-AND-COLUMN SYSTEMS

When actually drawing the beam-and-column systems used in residential applications, the detailer should remember that they are simply scaled-down versions of

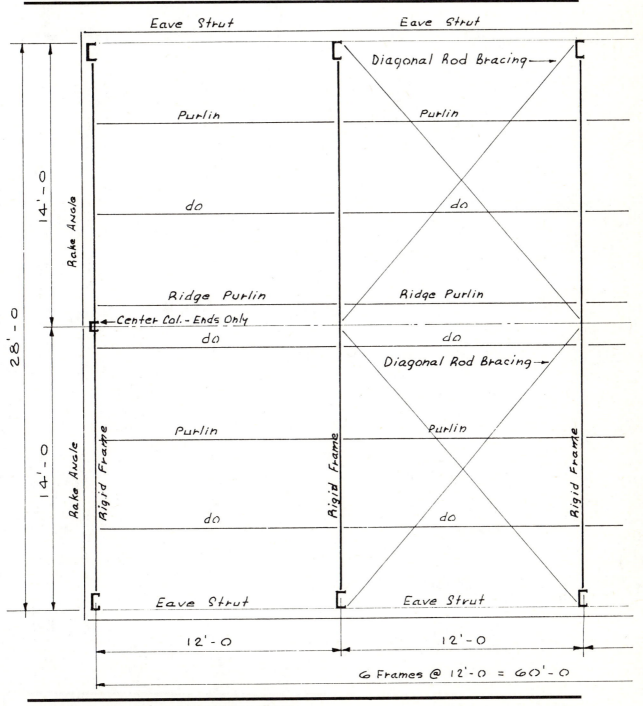

Figure 2-26 Partial framing plan for rigid-frame residence

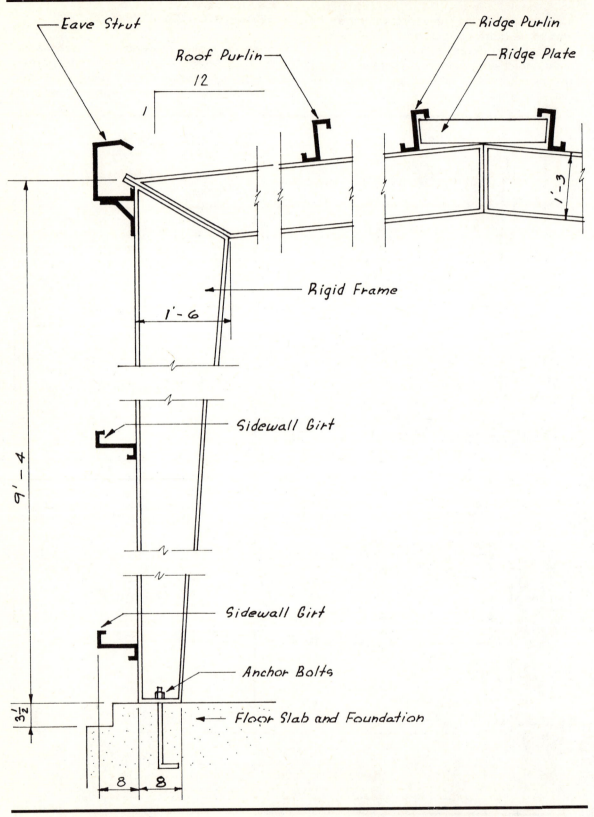

Figure 2·27 Rigid-frame detail

those found in commercial and industrial applications. The same practices used in developing drawings for the larger systems should be followed. Figure 2-22 provides the architect's display drawing of a beam-and-column residence. Figure 2-23 represents the framing plan (design drawing) for the structural system, with some of the accompanying details shown in Figure 2-24.

EXAMPLES OF PREFABRICATED METAL SYSTEMS

Prefabricated metal buildings and their components (rigid frames, girts, purlins, etc.) have been preengineered and highly standardized. Since they are very much a packaged product, the actual amount of drawing for such a system can be reduced somewhat. The partial residential floor plan in Figure 2-25 shows how the system is represented. Figures 2-26 and 2-27 illustrate the typical framing plan and details.

Metal building manufacturers provide manuals on their products, containing engineering data, detailed illustrations, and specifications of the various systems and their component parts. It is often unnecessary

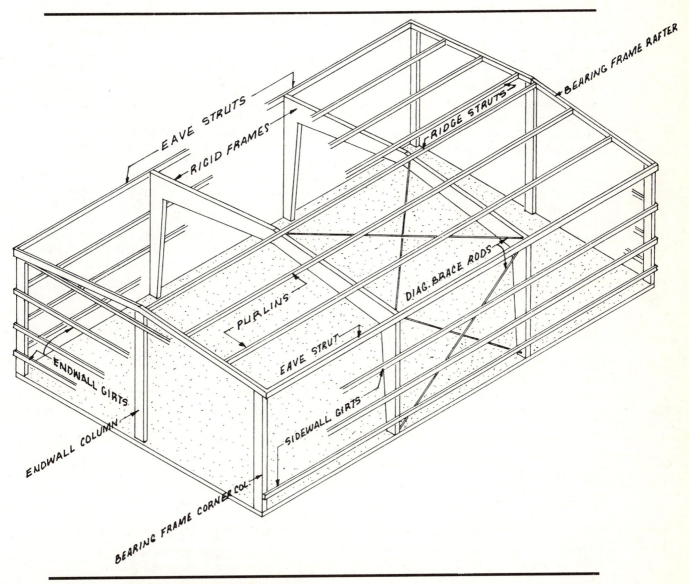

Figure 2·28 Typical prefabricated metal building frame

to detail all members, connections, and joints, since these may be assumed to be standard with the total package specified. The shop drawings for such structures may well be condensed when accompanied by an adequate bill of materials.

When detailing metal buildings (or simply the structural system) it is imperative to keep in mind that such systems should have diagonal wind bracing included. The braces (usually threaded steel rods with turnbuckles) should be placed at the corners of the structure and at such other points as are deemed necessary. Figure 2-28 illustrates a typical system with the various parts labeled with the common terminology. As can be clearly seen, stresses are highly concentrated when transmitted to the foundation system at the bases of the rigid frames, so attention must be given to the design of the foundation at these points. It is also important that the anchoring system for the rigid frames be carefully designed and detailed. Wind damage of severe proportion can result from improper anchorage of the rigid frames.

SUMMARY

As today's designers create residential structures, they are strongly influenced by the technological advances in the construction field. Cost factors, degree of permanence, ease and quickness of erection, and physical characteristics of the materials all have impact upon the designers' solutions to each building problem. Steel, with its inherent versatility and high degree of product standardization, is being seen more and more throughout the residential construction industry.

The older forms of beam-and-column, load-bearing, and long-span framing are being adapted for residential use. In addition, newer structural systems such as the rigid frame are finding acceptance by the home-building industry. Whether working with an old system revised or with a new system, the technician can ensure the correctness and completeness of the entire structural system by seeing that his work conforms to the quality and standards set by the industry. The transition from designer's thought to completed structure must depend upon the design drawings, detail drawings, shop drawings, and bills of materials created by the structural detailer.

Problems—Chapter 2

1. Fully dimension the W 12 × 50 beam shown in the illustration.

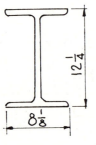

$12\frac{1}{4}$

$8\frac{1}{8}$

SECTION: W 12 × 50   LENGTH: 20'-4½
WEB THICKNESS: ⅜   FLANGE THICKNESS: ⅝

2. Prepare shop drawings for the column and plates shown.

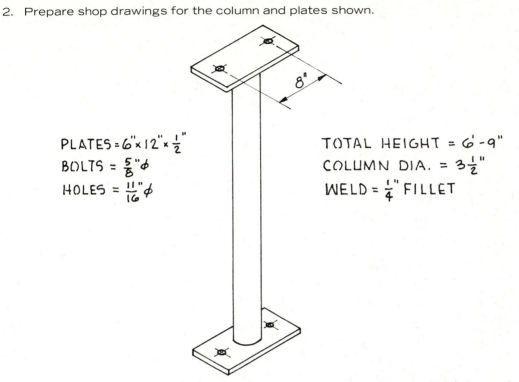

PLATES = 6"×12"×$\frac{1}{2}$"
BOLTS = $\frac{5}{8}$"φ
HOLES = $\frac{11}{16}$"φ

TOTAL HEIGHT = 6'-9"
COLUMN DIA. = 3$\frac{1}{2}$"
WELD = $\frac{1}{4}$" FILLET

3. Draw a plan view, details, and shop drawings for the basement girder and column system shown.

GIRDER SECTION : W 8 × 28
FLANGE THICKNESS : $\frac{7}{16}$
WEB THICKNESS : $\frac{5}{16}$
COLUMNS : STD. PIPE 3$\frac{1}{2}$"φ
TOP PLATES : 6 × 6 × $\frac{3}{8}$
BASE PLATES : 6 × 12 × $\frac{3}{8}$
WELDS : $\frac{1}{4}$" FILLET
BOLTS : $\frac{5}{8}$"φ HOLES : $\frac{11}{16}$"

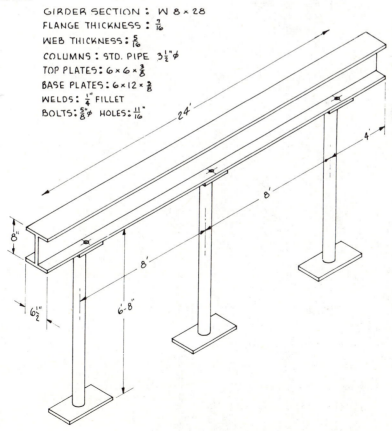

4. Prepare the shop drawing for the three beams shown, including the connecting angles.

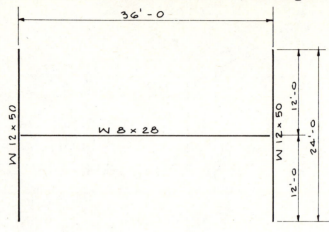

WEB AND FLANGE THICKNESSES GIVEN IN PROBLEMS #1 &#3.
ALL CONNECTING ANGLES TO BE 4 × 4 × 3/8 × 6.
ALL BOLTS TO BE 7/8 ∅ WITH 15/16 ∅ HOLES.

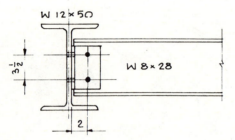

5. Draw details and shop drawings for the steel assembly shown in the illustration.

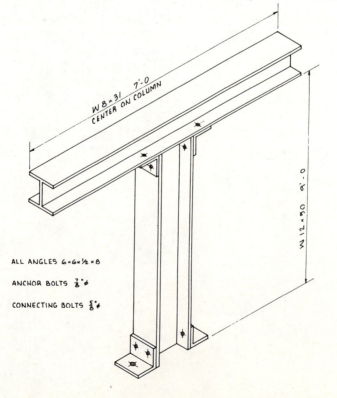

ALL ANGLES 6 × 6 × 1/2 × 8

ANCHOR BOLTS 7/8" ∅

CONNECTING BOLTS 5/8" ∅

6. Detail the end conditions for the beams shown.

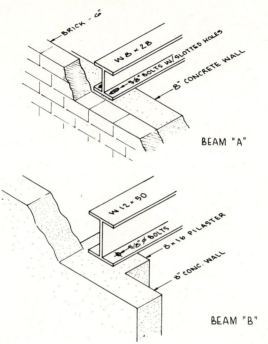

BEAM "A"

BEAM "B"

7. Create a bill of materials for the assembly drawn in problem 3.

After Studying This Chapter You Should Be Able To:

1. Name the three regular framing systems used in commercial construction.
2. Specify the type of framing system generally used for prefabricated (or preengineered) metal buildings.
3. Cite a number of structures which by their usage fall within the commercial category.
4. Explain why more care must be taken in designing *clean* joints and connections for commercial structures than is necessary with industrial structures.
5. Name the preferred style of lettering for structural drawings, and explain why it is chosen in preference to other styles.

INTRODUCTION

Commercial buildings have long relied heavily upon steel as a structural material, and, as mentioned in the chapter on the material, the high degree of standardization has resulted in excellent cost factors. When used in appropriate applications, steel components result in a rapid, economical method of construction.

Fire protection There are several efficient methods currently available for protecting steel members from fire damage. The members may be encased in masonry or concrete, wrapped with materials such as asbestos, or sprayed with fireproofing material. Proper treatment can overcome the susceptibility of steel to fire damage and can result in very satisfactory fire ratings.

chapter 3
steel structural systems- commercial

6. Give the general guidelines for the size of lettering used on structural drawings.
7. Identify the clearest method for forming fractions.
8. Explain what type of lines are used for: border lines, object lines, invisible lines, dimension and extension lines, match and gage lines, and center lines.
9. Describe cutting-plane, short-break, and long-break lines.
10. Describe the proper proportions, type, and size preferred for arrowheads on structural drawings.
11. List some of the various types of shop bolt, field bolt, and rivet symbols.
12. Show how a welding symbol is constructed, and explain what its basic parts indicate.

Structural forms The extremely wide range of standardized structural steel sizes and shapes available permits commercial structures to be designed with infinite architectural variety. Practically any form of structural system can be created with steel components of standard production, and special structural forms can be easily created from composite units built up from smaller standard member shapes.

Framing types Commercial buildings make use of structural steel in wall-bearing, beam-and-column, and long-span framing systems. Present-day steel bar joists and trusses permit long, clear spans, resulting in buildings with large, open areas of unobstructed floor space. The relatively recent domes, space frames, and suspension systems provide even greater areas of clear, usable space. They often preclude the need for heavy, permanent walls.

Remodeling   By their very nature, commercial buildings must fit the purpose for which they are to be used. Flexibility of space for commercial purposes has become increasingly important and structural systems which permit light, non-load-bearing walls offer the potential of easy future remodeling.

Exposed structural members   Contemporary architecture, in the form of modern commercial buildings, very clearly indicates that exposed structural systems can provide desirable architectural design features. Such exposed structural elements can easily serve a decorative as well as a functional purpose. The self-oxidizing steels provide an extremely attractive appearance. Beam-and-column structural designs lend themselves particularly well to the exposed-structure format and are very much in evidence in the present construction scene.

Prefabs   Prefabricated (or preengineered) metal (steel) building systems are very much a part of current commercial construction and are a form of long-span framing. As with residential applications, these systems may provide the structural system alone or the entire basic building complete with roof and wall panels. Since many preengineered metal building systems are based upon some form of rigid frame, the interior space can be relatively (or totally) free from the encroachment of columns.

Remodeling potential of metal buildings is quite good, and even the exterior walls permit ease of creating new openings and expanding interior room sizes. The metal panels can be easily removed, and the only really fixed components at floor level are the legs of the rigid frames.

Metal buildings, once limited primarily to small shops, manufacturing facilities, and warehouse usage, are now functioning in numerous "commercial" applications. Churches, stores, offices, restaurants, and schools are currently being constructed with metal building components and complete metal building systems.

COMMERCIAL BUILDINGS

The term *commercial building* as used herein is meant to encompass that broad spectrum of structures spanning between residential buildings (providing actual dwelling) and the buildings associated with industry (for storage, processing, and manufacturing). Taken in this context, then, the commercial classification can include schools, churches, financial institutions, wholesale and retail sales establishments, service facilities, recreational and resort facilities, public and governmental buildings, hospitals and clinics, and office structures.

Certainly, the type of usage demanded for commercial buildings dictates that appearance is an important design factor. In those designs where the structural members are to be covered with finish materials, the detailer must design connections and joints to facilitate the application of the finish materials. Commercial buildings where the structural members are to be exposed to view demand even greater attention to providing clean, attractive joints and connections.

ADEQUATE DRAWINGS

The basic guidelines for the work of the structural detailer cannot be overemphasized, whether the work is of residential, commercial, or industrial nature. The detailer must provide sufficient details for the estimation, fabrication, shipping, and erection of the structural system. Such details must be complete, concise, and easy to comprehend. Common sense rules out the inclusion of minutiae and trivia which serve only to confuse the issue. With background information provided in the two previous chapters, this chapter examines steel detailing practices, particularly as they relate to the commercial building types.

The actual structural drawings must depend, for their validity, upon the accuracy of the detailer and his use of accepted practices and conventions. Much less effort is needed to explain an item if conventional detailing methods are used. The men in the shops and in the field have come to rely upon certain reoccurring (generally accepted) practices and symbols as a fast aid to understanding and interpreting the drawings. A number of these conventions have been mentioned previously, but the coverage in this chapter is more complete.

LETTERING ON DRAWINGS

Symbols and drawing practices certainly provide a means to clearer understanding of structural systems and details, but there is still a wealth of important information provided on drawings via the printed word. Dimensions are given by numerals printed by the detailer; marks and numbers provide a clear-cut designation of structural members for fabrication, shipping, and erection; and the notes and call-outs hand-lettered by the detailer provide additional information of equal importance. Lettering must, therefore, be clear and legible enough to absolutely ensure that the various characters will not be misread. The change in meaning brought about by a misplaced or poorly formed letter or numeral can have drastic and far-reaching consequences.

Some firms have adopted such lettering devices as Leroy and Wrico and Varityper in their efforts to ensure readable drawings, but good freehand lettering continues to be the predominant method of presenting the information. Architectural firms have maintained a personal quality to their lettering, while engineering firms have moved more toward a standard format. The advent of microfilming brought added impetus toward a standardized system for hand lettering.

Lettering on structural drawings (with a few variations among firms and areas) has been generally standardized and is now based upon the single-stroke lettering shown in Figure 3-1. The inclined lettering is more popular than the vertical because it is easier to achieve a uniform appearance with it. The simplicity of this inclined lettering style and the rapidity of its execution have helped to make it the most-used style in the structural field.

Lightly drawn guidelines should be used for lettering on drawings to ensure consistency of the lettering and to provide adequate horizontal alignment. Slope guide-lines may be necessary for the beginning detailer but are not generally used by the experienced structural detailer in executing inclined lettering.

Notes on structural drawings usually consist of lowercase (small) and uppercase (capital) letters. Titles and title block information may be given in all uppercase letters. To be easily read, uppercase letters and numerals should be a minimum height of $5/32$ in., with a proportionate decrease in size for lowercase letters.

Possibly no item of information can be so easily misinterpreted if poorly formed as the fractions necessary to call out thicknesses, spaces, and dimensions. Fractions should be formed with bars separating the two numerals (numerator and denominator). Fractions are often seen with the diagonal (sloping) bar dividing the numerals. This practice reduces the overall height of the fraction but permits improperly formed bars to be misinterpreted for the numeral 1. The safest method is to rely solely upon the horizontal division bar.

Students should give ample time to practicing lettering to develop both proficiency and speed. Firms vary in their requirements, and the detailer may be lettering with pencil (or plastic leads) or ink on drafting vellum, cloth, or Mylar. Practice is the key to success for lettering in any medium. Most firms have a well-detailed set of drafting room standards which carefully explain their own requirements for letter styles, size, dimensioning practices, etc.

LINES AND LINE QUALITY

It is not unusual for drafting instructors to hear from industrial employers that if the instructor teaches the student to lay down a good, heavy line consistently, half of the training process will have been accomplished. This, of course, is a gross oversimplification of training for a highly technical field. It does point, however, and rightly so, to the fact that the quality of the actual drawing is important.

Light drawings and inconsistent line quality become real problems when drawings

ABCDEF GHIJKLMN

SINGLE - STROKE VERTICAL

SLOPE   GUIDELINES

ABCDEF

GHIJKLMNOP

QRSTUVWXY

Z&123456789

1'-8½   40'-9   7⅞   6'-5

SINGLE - STROKE INCLINED - UPPER CASE

SLOPE   GUIDELINES

abcde

fghijklmno

pqrstuvwx

yz

SINGLE-STROKE INCLINED - LOWER CASE

Figure 3·1 Lettering styles

are sent through the reproduction process. Regardless of how long a detailer has spent on a given set of drawings, his efforts are practically nullified if the original drawings fail to reproduce clearly. Each firm includes requirements for line symbolism and quality in its drafting standards to provide and promote uniformity of all work produced within the firm. Figure 3-2 indicates the basic lines necessary for most structural detailing.

The example in Figure 3-3 shows how various lines appear in actual usage and explains their particular functions on the structural drawing. Notice that the difference in weight of the various lines helps (rather than confuses) the viewer in interpreting what is being presented.

Border lines are the heaviest lines used on the drawing. The title block is generally formed with borderline thickness. These lines are quite heavy to set them apart from the drawings themselves. Many firms now use preprinted borders and title blocks, and most of these commercially prepared drawing sheets conform to the borderline weights recommended herein.

Object lines are less pronounced than border lines but are still the heaviest of the

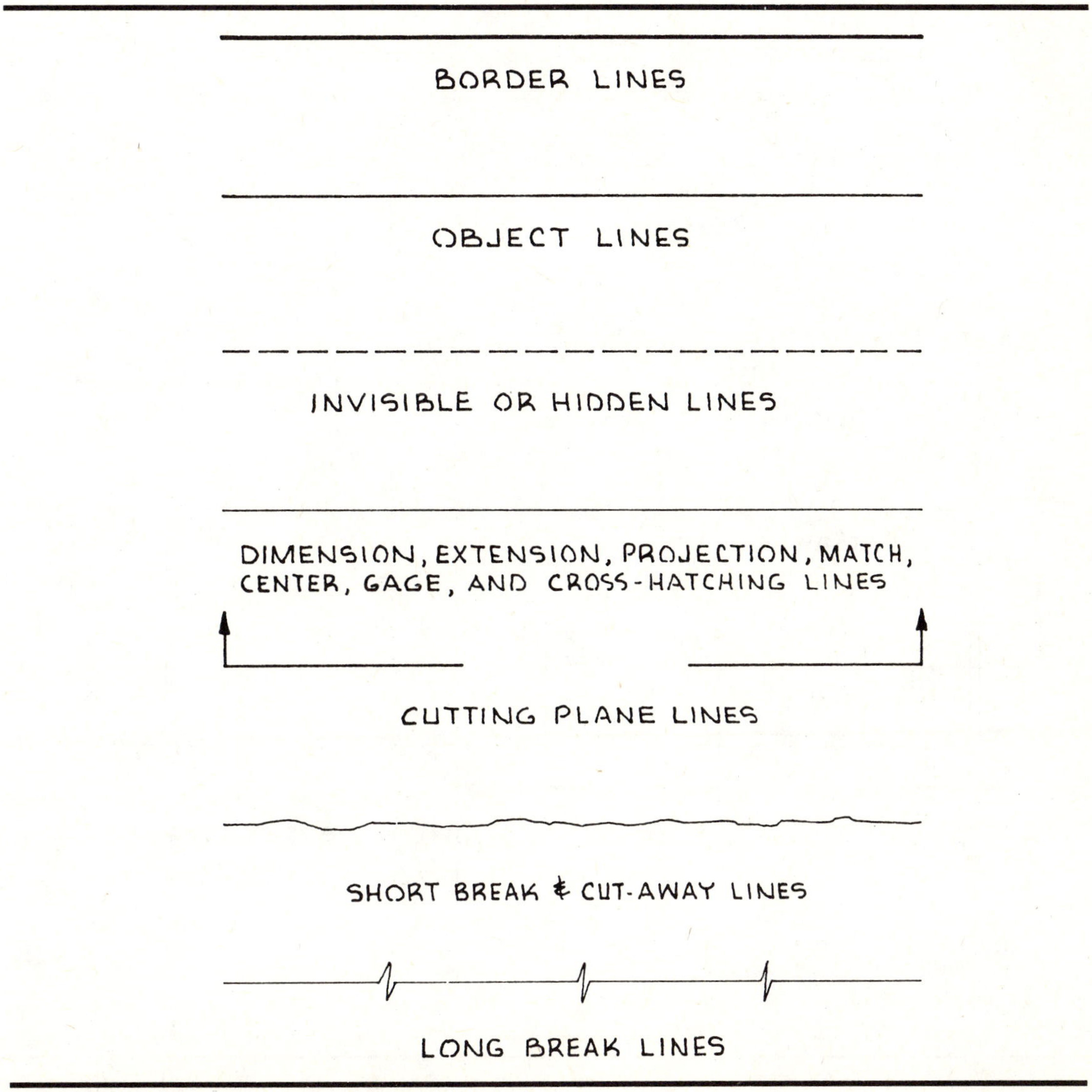

BORDER LINES

OBJECT LINES

INVISIBLE OR HIDDEN LINES

DIMENSION, EXTENSION, PROJECTION, MATCH, CENTER, GAGE, AND CROSS-HATCHING LINES

CUTTING PLANE LINES

SHORT BREAK & CUT-AWAY LINES

LONG BREAK LINES

Figure 3·2 Lines used for structural details

actual drawing lines. Since many extension lines, dimensions, and call-outs are applied to the basic object drawing, it is imperative that the actual object be shown clearly to avoid confusion. The object must stand out clearly from the supportive information.

Invisible (or hidden) lines are used to show

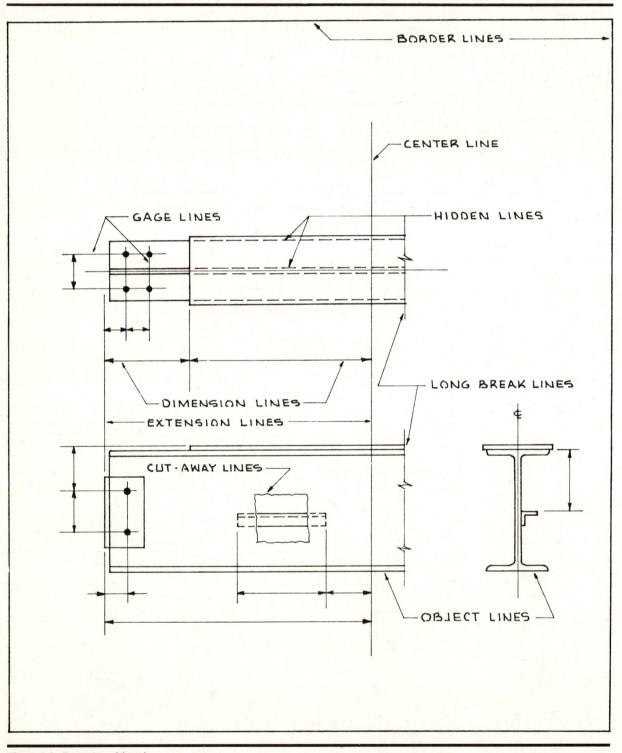

Figure 3·3 Examples of lines in use

hidden edges or features of the object which do not appear in the regular view. These lines are slightly lighter in weight than the regular object lines. They are formed with a series of uniform dashes approximately twice as long as the spaces between them. As a rule of thumb the dashes should be $\frac{1}{8}$ in. long, with $\frac{1}{16}$-in. spaces separating them. The detailer need not scale the dashes and spaces but should be able to create consistently formed lines with little variance in the weight and size.

One particular type and weight of line is called upon to serve a number of important purposes. It is a solid unbroken line, lighter in weight than object or invisible lines. It is used for dimension, extension, projection, match, gage, center, and cross-hatching lines.

Cutting-plane lines are used to show the location of an imaginary cutting plane, for the purpose of drawing sectional views. These lines are solid and equal in weight to the object lines. The cutting-plane lines

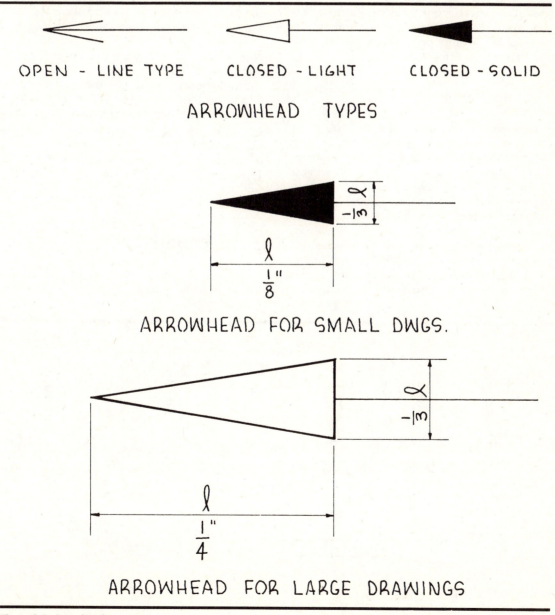

Figure 3·4 Arrowheads

have arrowheads to indicate the direction of the section being viewed.

Short breaks and cutaway sections utilize a ragged, wavering line equal in weight to the dimension lines. Long breaks, on the other hand, are indicated by straight lines of the same weight, broken with the traditional zigzag indication associated with the break.

It should become obvious to the beginning student that a great amount of the effectiveness of structural drawing depends upon the proper choice of conventional lines. Departure from the long-established line practices must result in confusion and ineffectiveness in conveying the necessary structural information.

ARROWHEADS

Arrowheads are a necessary part of drawing, and most firms specify very clearly their requirements for the construction of arrowheads on structural drawings. In general, all arrowheads on a single drawing will be of uniform size. The size selected is, of course, chosen to be appropriate to the drawing itself. Small drawings may very well call for 1/8-in. arrowheads, with the size ranging up to 3/16 or 1/4 in. on large drawings. Arrowheads are most often proportioned so that the length is

three times the width. Although some firms permit the open or single-line type, most detailers will be called upon to use the solid type. Both types of arrowhead appear in Figure 3-4.

BOLT SYMBOLS

Symbols for bolts are generally designated for two types of application. One group consists of shop bolts (those attached in the fabricating shop), and the other group represents field bolts (those used in on-site erection). The symbols in Figure 3-5 account for both groups and show the designation for each in elevation and section.

Notice that symbols are provided for hexagonal and square head bolts and for countersunk head bolts with the head on the near side (N.S.) and far side (F.S.). One of the common methods of call-out consists of encircling the bolt in elevation and giving number, type (H.S.B.—high strength; csk—countersunk head; and P.B.—other types), diameter, and length. Additional description may call out the type of nut and washer also.

RIVET SYMBOLS

The method of showing rivets follows a pattern similar to bolt symbols and has

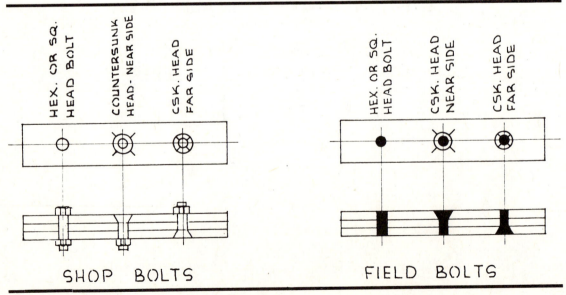

Figure 3·5 Bolt symbols

varying symbols for shop rivets and field rivets. There are many more commonly used symbols for rivets because of the larger variety of rivets and methods available. The symbols in Figure 3-6 are for the most common structural applications. Notice that the symbols are standardized for features appearing on the near side, far side, and both sides.

Rivets, like bolts, may be called out by encircling them and adding individual notes, by providing a schedule of fasteners for the various joints and connections, or by using call-outs combined with general fastening notes.

WELDING SYMBOLS

Welding symbols have been standardized to a high degree and provide a quick, graphic representation of the type of weld required for any given location. Each com-

plete welding symbol consists of three parts, with a fourth part added when necessary to furnish additional information.

The common symbol consists of an arrow pointing to the joint where the weld is to occur, a reference line to carry lettering for dimensioning, and the basic weld symbol indicating the type of weld to be used. The tail on the reference line is added only when additional information is necessary regarding processes or references. Figure 3-7 shows the typical construction of the completed symbol and provides many of the basic and supplementary weld symbols used to indicate weld types.

When constructing weld symbols, the symbols with vertical legs should be drawn with the vertical leg to the left. As seen in the illustration, a field weld is indicated by the addition of a solid dot at the point of intersection of the arrow and the reference line. Some basic weld and joint types were

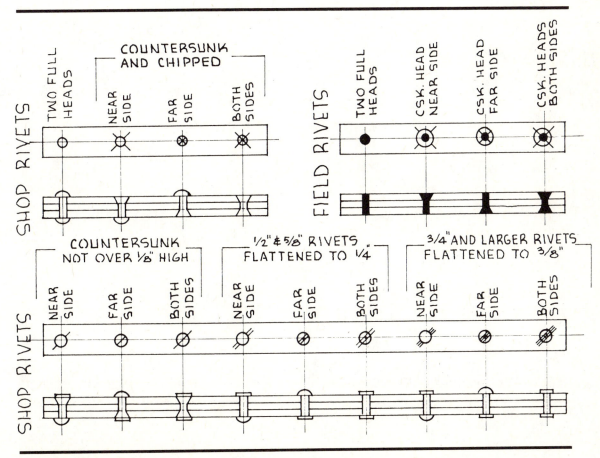

Figure 3·6 Rivet symbols

illustrated in Figures 1-16 and 1-17 of Chapter 1. The student is advised to review those figures to familiarize himself with the intent of the welding symbols.

EXAMPLES OF WALL-BEARING SYSTEMS

Commercial structures often utilize the

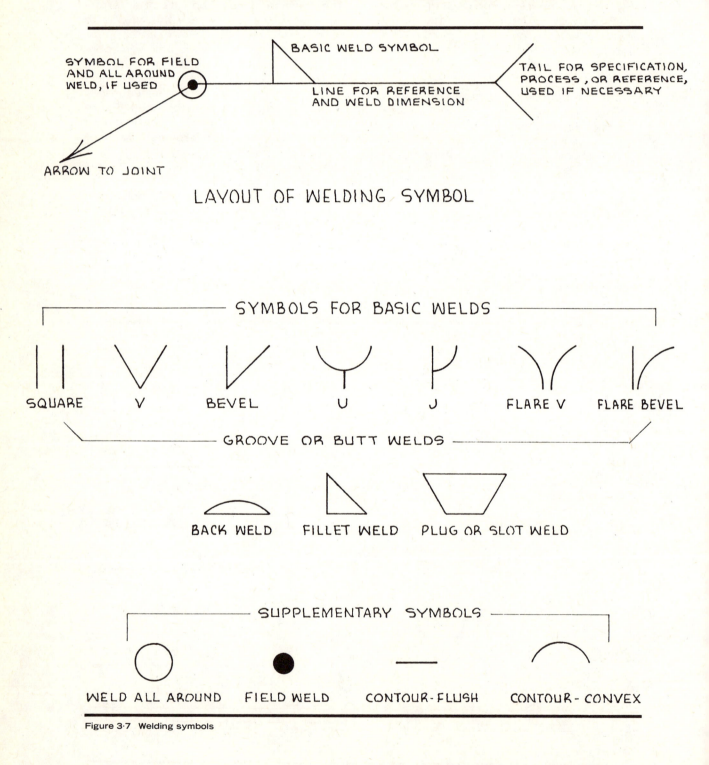

SYMBOL FOR FIELD AND ALL AROUND WELD, IF USED

BASIC WELD SYMBOL

TAIL FOR SPECIFICATION, PROCESS, OR REFERENCE, USED IF NECESSARY

LINE FOR REFERENCE AND WELD DIMENSION

ARROW TO JOINT

LAYOUT OF WELDING SYMBOL

SYMBOLS FOR BASIC WELDS

SQUARE   V   BEVEL   U   J   FLARE V   FLARE BEVEL

GROOVE OR BUTT WELDS

BACK WELD   FILLET WELD   PLUG OR SLOT WELD

SUPPLEMENTARY SYMBOLS

WELD ALL AROUND   FIELD WELD   CONTOUR-FLUSH   CONTOUR-CONVEX

Figure 3-7 Welding symbols

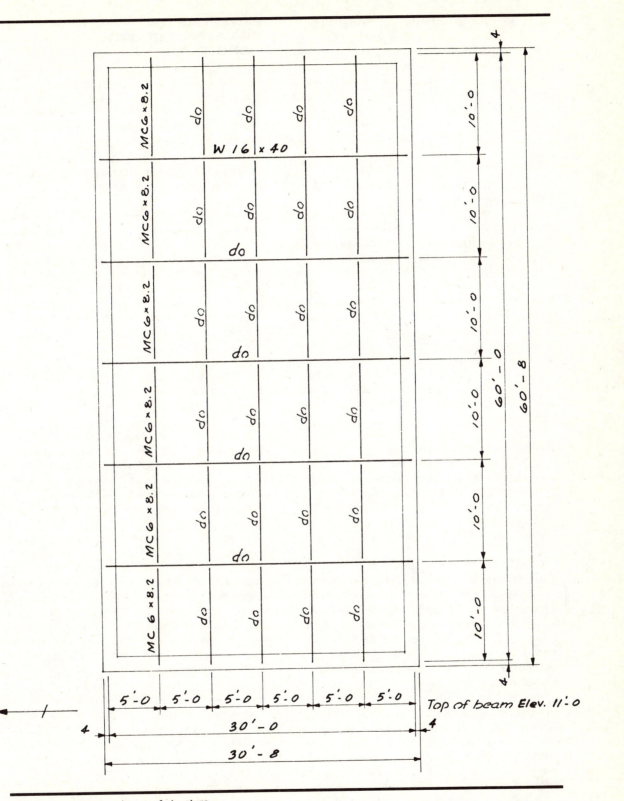

Figure 3·8   Design drawing–roof structure

wall-bearing type of framing. Steel beams or light bar joists span between masonry walls capable of supporting the load imposed by the framing members. This particular type of construction is most feasible for single (or, at the most, triple) story structures. Many of today's commercial buildings are single story, and such a system is satisfactory for such an application.

The commercial building shown in Figure 3-8 is typical of the small commercial structure and has concrete-block walls with 4-in. brick veneer on the exterior. Standard steel beams form the roof structure, with steel purlins and steel decking completing the roof construction. In this instance, a suspended ceiling is hung from the roof structure. The wall-bearing system is used throughout the building.

The framing plan in Figure 3-8 indicates the structural steel necessary for construction of the project. In this particular case, a typical beam is used repetitively, and a single set of details is applicable for purlins as well.

Some of the shop drawings for the structure are indicated in Figure 3-9 and represent what would ordinarily be required for this type of structure. This particular building system relies upon field bolting for the joints and connections. The symbols used are similar to those given in Figure 3-5.

The purlins for this structure are small steel channels bolted to the major beams. Notice that the channels face the same direction. They could face different directions at the back (left) of the structure. This change in direction of face of the members would mean that these units would be fabricated and erected similar but opposite to the bulk of the purlins. Some detailers might choose to call them out as being *similar but reversed*. Many instances occur in commercial steel structures calling for right- or left-hand connectors, beams, etc. In such cases, details are drawn for both and labeled with the same general number followed by an R or L. A single connector may be shown with the note that the other is similar but reversed, or opposite hand.

EXAMPLES OF BEAM-AND-COLUMN CONSTRUCTION

Commercial structures often make use of the beam-and-column system of framing, particularly in multistory design. This framing system facilitates erection of high structures, and with the freestanding structural skeleton, only lightweight construction is required for enclosure (curtain walls with non-load-bearing components).

Multistory framing Figure 3-10 represents a partial framing plan for a three-story commercial office building. The bays in such a structure usually repeat for each tier, but the various floors should be detailed and drawn separately to avoid confusion and possible deletion of items. In addition, member sizes often change from one tier to the next, even though the basic bay dimensions remain constant.

The joints and connections in this structure combine shop and field bolting and welding. The details in Figure 3-11 show some of the necessary drawings for the structural system. Notice that to facilitate positioning and temporary support of members for field welding, connectors often have a few field bolts included. Such a practice is extremely helpful during erection and occurs frequently.

The welding symbols used on the sample details in Figure 3-11 are based upon the basic symbols shown in Figure 3-7 and provide the information needed in the shop and the field.

Columns often run more than one tier in continuous fashion. The details shown in Figure 3-12 indicate that, in this structure, columns run unbroken for two tiers (stories) with the top-story columns spliced to the columns for the first two floors. This practice permits structural continuity, and the size of columns on the top floor (supporting roof load only, rather than dead and live loads of roof and floors) can be reduced somewhat.

The column system shown in Figure 3-12 is of particular advantage in this installation because the third story is smaller in area

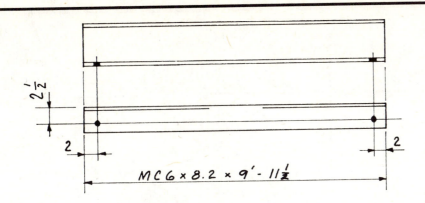

ROOF PURLIN

Note. All holes $\frac{11}{16}$ except where noted otherwise.

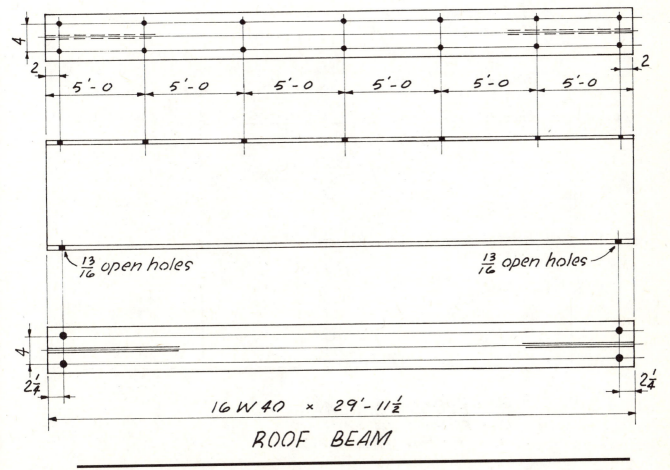

ROOF BEAM

Figure 3·9 Drawings for purlins and beams

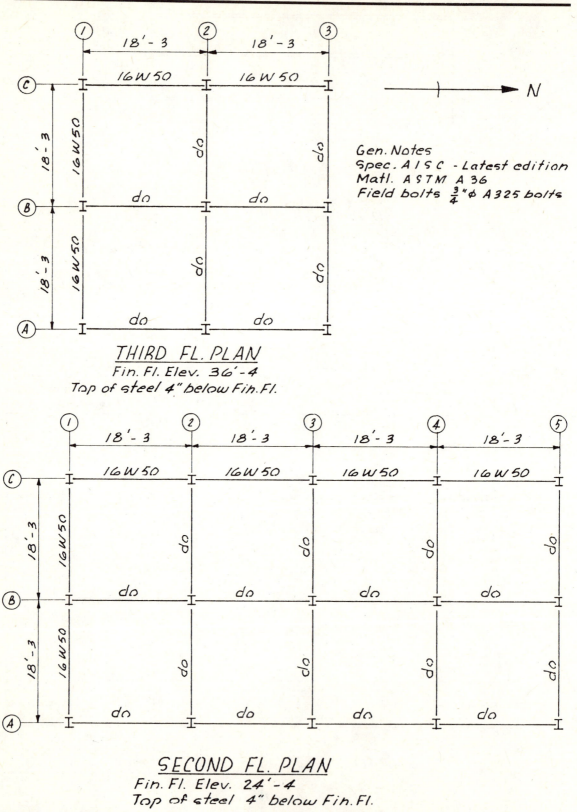

Gen. Notes
Spec. A I S C - Latest edition
Matl. ASTM A 36
Field bolts $\frac{3}{4}$"ϕ A325 bolts

THIRD FL. PLAN
Fin. Fl. Elev. 36'-4
Top of steel 4" below Fin. Fl.

SECOND FL. PLAN
Fin. Fl. Elev. 24'-4
Top of steel 4" below Fin. Fl.

Figure 3·10 Floor framing plans

than the other floors; thus the main first- and second-tier columns can all be identical sections, with the third tier being an add-on type of unit for the portion of the structure where it is required.

Single-story framing   The single-story structure can make good use of beam-and-column framing. The single-story structure in Figure 3-13 uses such a framing system. This particular building has a wide roof

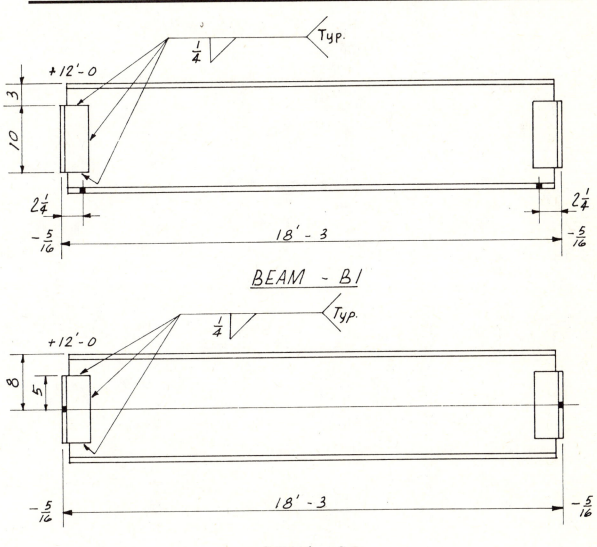

Figure 3·11 Framed beam details

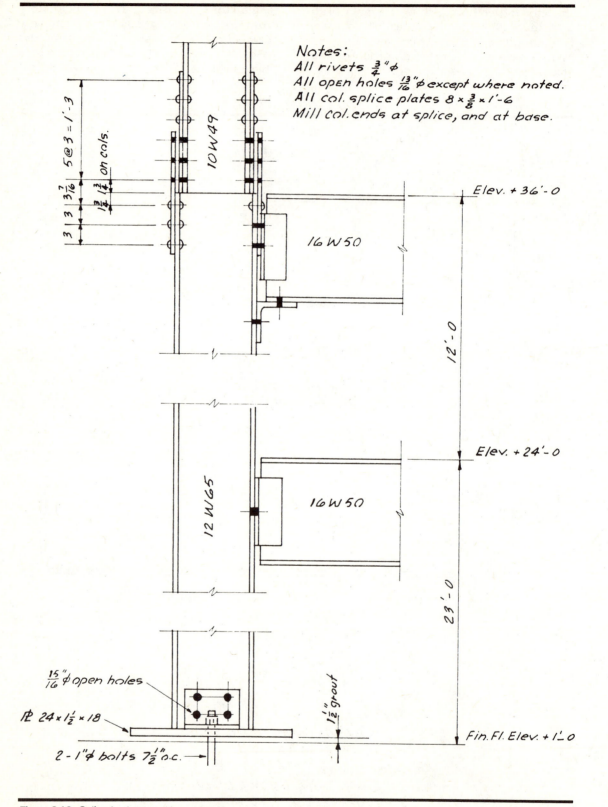

Notes:
All rivets $\frac{3}{4}"\phi$
All open holes $\frac{13}{16}"\phi$ except where noted.
All col. splice plates $8 \times \frac{3}{8} \times 1'-6$
Mill col. ends at splice, and at base.

Figure 3·12 Spliced column and beam detailing

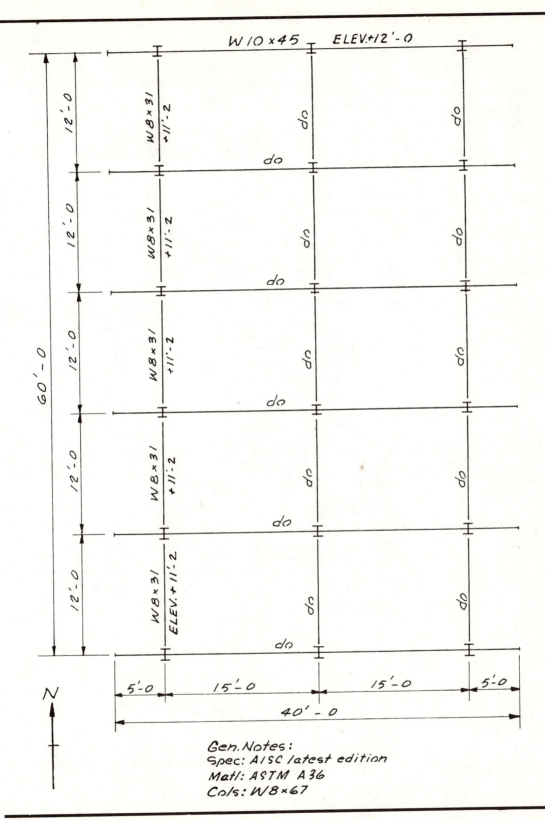

Figure 3·13 Beam-and-column framing plan

overhang on the two sides, so the roof beams actually continue over the exterior columns (cantilever) to support the roof and form the overhang.

A detail for the beams and the columns to which they attach is shown in Figure 3-14. The shop-fabricated units in this case are shop-riveted, and the field attachment is done by field bolting. Many of the smaller construction contractors are not equipped to do riveting, and distance and location are often factors precluding field riveting. In these instances, field welding or bolting is to be preferred for quick and economical erection.

Notice in these details that a *seated* con-

nection has been used to support and position the beams prior to and during the application of the field bolts. These seated connections occur for the longitudinal beams spanning between the main beams of the system.

EXAMPLES OF LONG-SPAN FRAMING

The use of long-span framing is widespread, as mentioned previously, and eliminates columns and walls obstructing the interior building space. The example cited in Figure 3-15 is for a single-story commercial structure using load-bearing (brick) walls and long-span bar joists. The bar

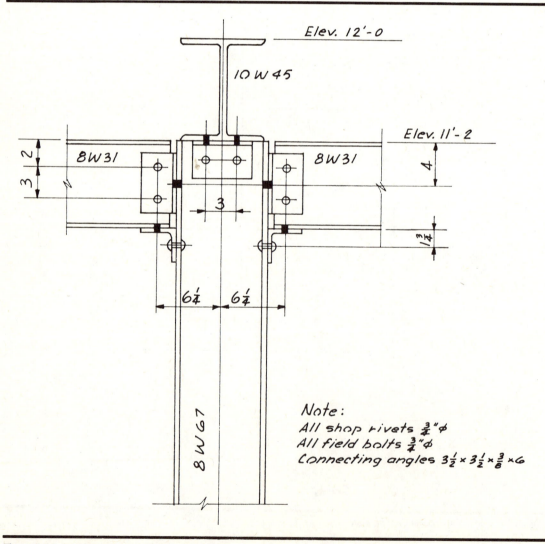

Figure 3·14 Interior column and beam detail

joists shown in the drawing (framing plan) are of the heavy series. These joist units are placed close enough together to permit corrugated steel decking to span the distance without purlins.

The details shown in the figure pertain to the end conditions and attachment details

for the bar joist units. Since the bar joists are a standard product, shop drawings are not furnished for them. They are, instead, specified by the standard nomenclature for the desired units.

Steel trusses provide another easy means to achieve large, open space for commer-

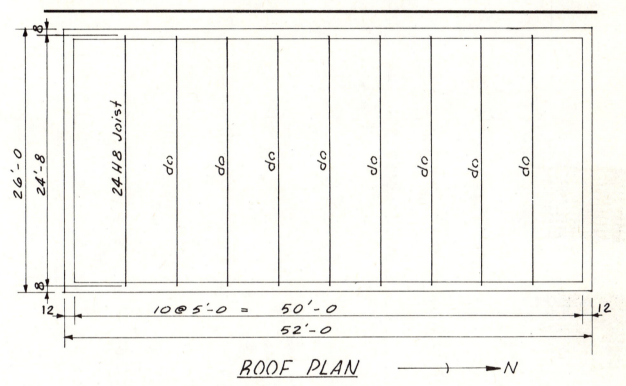

ROOF PLAN ⟶ N

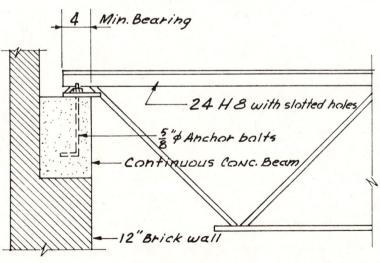

END DETAIL

Figure 3·15 Framing plan and bar joist end detail

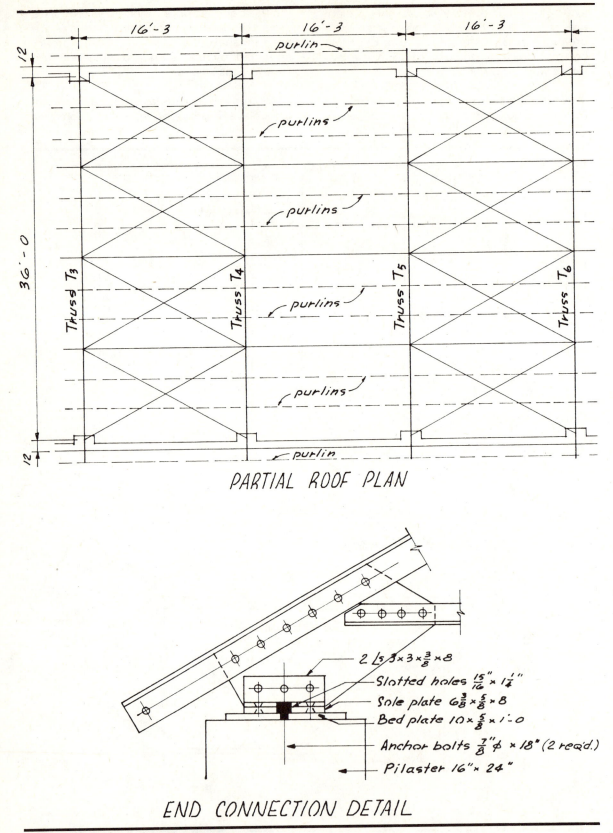

PARTIAL ROOF PLAN

END CONNECTION DETAIL

Figure 3·16 Partial plan and truss end detail

cial purposes. The building which is represented by the partial framing plan and details of Figure 3-16 uses steel trusses for its long-span framing system. Notice that integral pilasters have been constructed with the wall construction to counteract the heavy concentration of loads at the points of truss end-bearing. The pilasters, rather than the wall itself, absorb and

transmit the stresses from the trusses to the foundation.

The longitudinal bracing between roof trusses is very much recommended, as mentioned in a previous chapter, to reduce the effects of wind load and its accompanying tendency to rack (twist) the structure. Some of the bracing members are de-

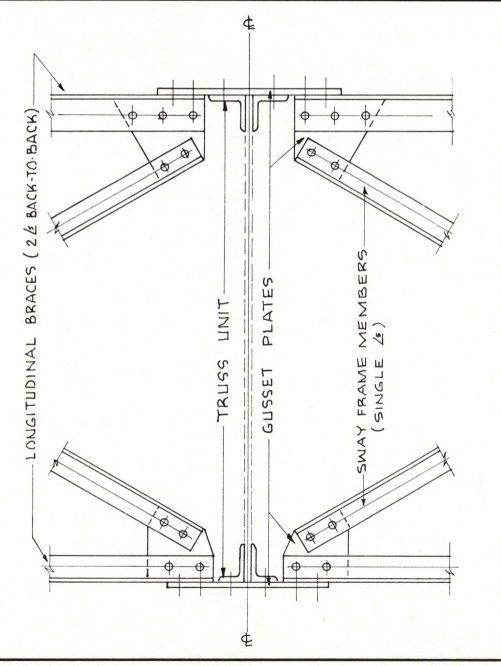

Figure 3-17 Detail of longitudinal braces and sway frame

tailed in Figure 3-17. These particular structural members are steel angles which are attached to the trusses with gusset plates and field bolts.

TRUSS DETAILS

The structural detailer may be called upon to create the drawings necessary for the shop fabrication of steel roof trusses. These drawings must be complete to the extent of specifying overall dimensions, component sizes, roof pitch, joints, and fasteners. Generally, the overall dimensions and pitch are known, since this information is an integral part of the building design. The following represents one method of developing the necessary drawings for a roof truss. In this example only a few truss components are detailed individually for actual shop fabrication, but in reality each component and joint would be completed in shop drawing form.

The truss in this instance would be detailed assuming that the span, rise, and structural design have been prepared by an engineer and sent in rough form to the detailer.

The preliminary layout is done by establishing the member positions with their *gage lines* (the center lines of holes for fasteners). The various steel manuals provide recommended gage-line spacings for the different steel shapes. The completed gage-line layout is identical to the final configuration of the truss and establishes the *slopes* required for the separate members. This sample truss is to be fabricated from steel angles back to back, with riveted fastening through steel gusset plates. The ends of the truss are to be provided with *sole plates* (with slotted holes) for attachment to the bearing plates on top of the wall pilasters.

The partial drawing of the truss in Figure 3-18 shows the appearance of the truss laid out in gage lines. Notice that top and bottom chords, as well as the webs (diagonal members), are all indicated by the gage lines.

Once the basic truss has been laid out with gage lines, the detailer proceeds to complete the drawing showing the actual members, gussets, and fasteners. The truss should be shown in front view, with top views of top and bottom chords in-

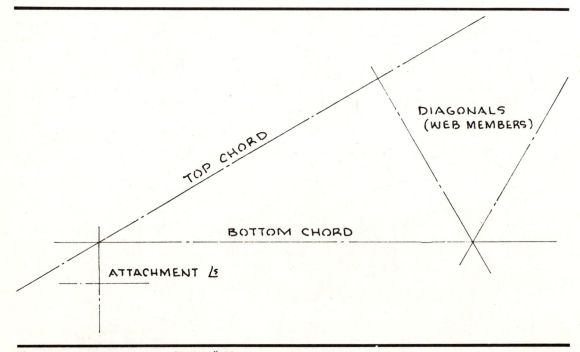

Figure 3·18 Partial truss layout with gage lines

cluded. The partial drawing in Figure 3-19 represents the finished truss drawing. Notice that holes are shown and specified to receive bolts from roof purlins. It is common practice to indicate slopes or bevels of diagonal members rather than angles. The symbols on this drawing for the bevels are the small triangles, giving horizontal/vertical rise ratios in the form of the proportions of the sides of the right triangle.

One of the details of the truss appears in Figure 3-20. The partial detail in the figure is for the ridge (or peak) and bottom chord of the truss. To facilitate shipping, the truss is built in two sections which are joined with field bolts at the job site. The detail indicates that the members at the ridge are joined by this method. Notice that the members on the right of the center are fastened to the gusset plate in the shop, while those on the left side are field-fastened. The two sections of truss are field-joined at the bottom chord by three bolts at the gusset plate to the left of the center

line of the truss and the same at the gusset to the right, since the truss is symmetrical about the center line. The center section of the bottom chord is thus totally attached on the job, rather than in the shop.

Each of the joints of the truss should be clearly detailed in the manner of the examples.

A bill of materials can be prepared for the truss from the completed drawings, listing each item necessary for the fabrication and erection of the truss. The completeness of the truss drawings and details makes it easy to arrive at an accurate bill of materials and facilitates quick and accurate fabrication with nothing left to guesswork.

The detail in Figure 3-21 indicates how the same truss might be fabricated by welding. The detail is at the ridge and bottom chord, and the member sizes are the same. One member has been added as a vertical bracing member at the center of the truss.

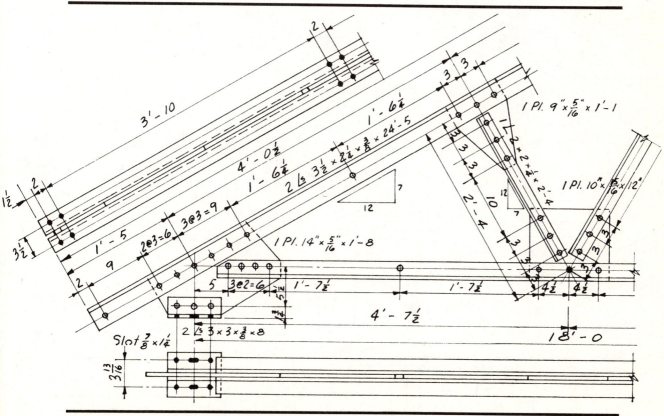

Figure 3·19 End portion of detailed truss drawing

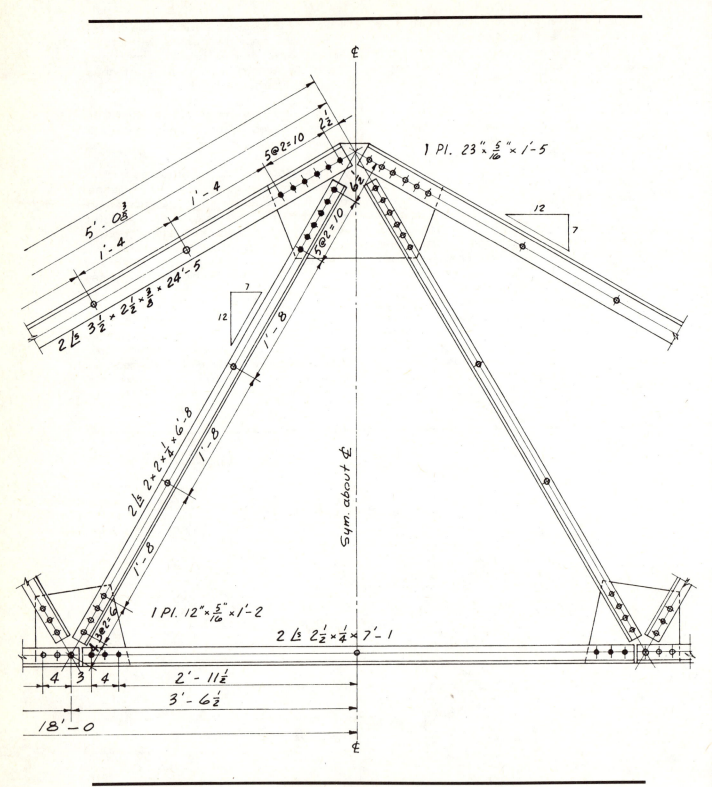

Figure 3·20 Center portion of detailed truss drawing

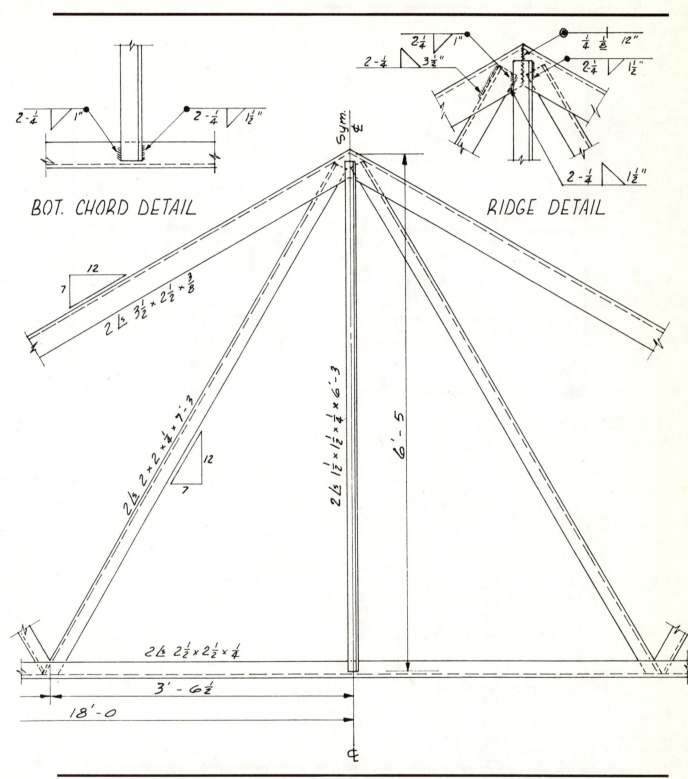

Figure 3·21 Center portion of detailed truss drawing and details

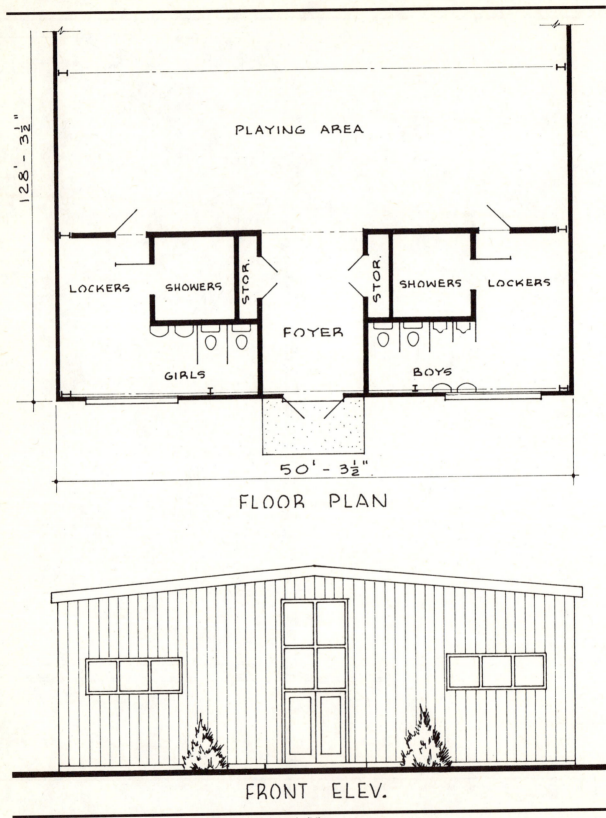

128' - 3½"

PLAYING AREA

LOCKERS   SHOWERS   STOR.   STOR.   SHOWERS   LOCKERS

FOYER

GIRLS   BOYS

50' - 3½"

FLOOR PLAN

FRONT ELEV.

Figure 3·22 Preliminary drawing for prefabricated metal gym building

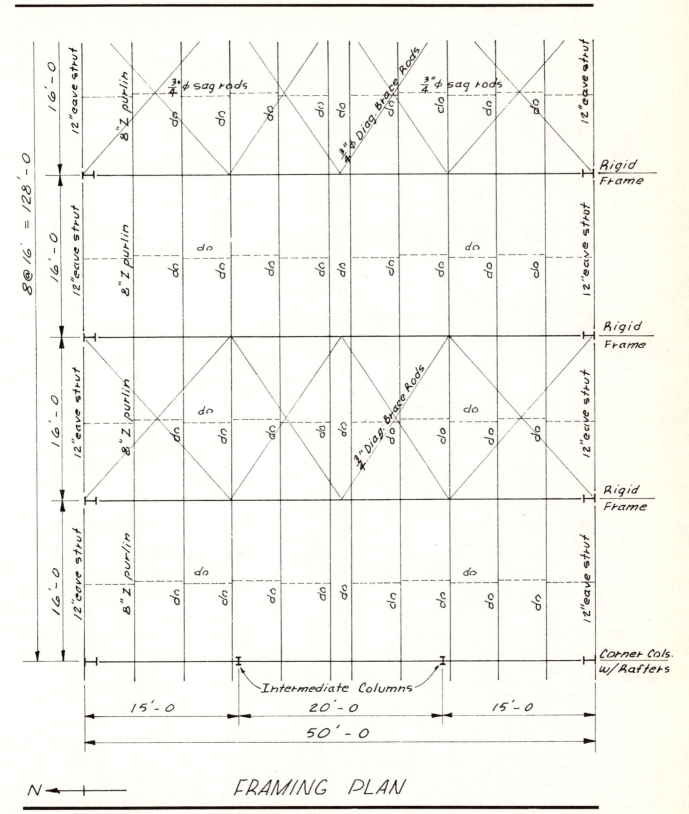

FRAMING PLAN

Figure 3·23 Partial framing plan for prefabricated metal gym building

The actual detail drawing shows the joint with the members on the front side removed for the sake of avoiding confusion, but the welding symbols indicate the welding required in the two different planes required for the back-to-back members. The marking of the actual welds with *hatching* is not necessary with the weld symbols but is shown here to show the reader more clearly what is meant by the symbols.

PREFABRICATED METAL BUILDINGS

The preengineered metal building types have many commercial applications. The

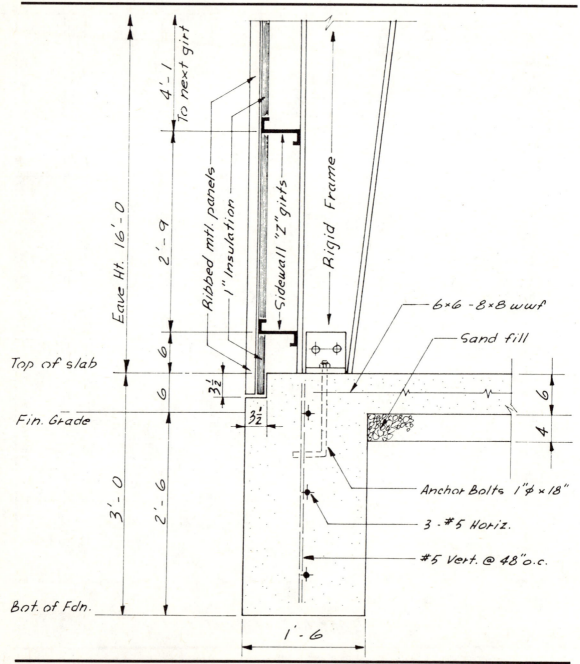

Figure 3·24 Rigid frame base detail

building shown in Figure 3-22 is an elementary school gymnasium and all-purpose room. The entire building is designed around a standard metal building package. The structural system is made up of light steel rigid frames with channel girts and purlins. The exterior panels for roof and walls are ribbed steel attached to the girts and purlins.

A partial framing plan appears in Figure

3-23. Notice that the legs of the rigid frames are carefully located with center lines and are fully dimensioned.

One of the typical details for the building appears in Figure 3-24. The detail shows the metal skin (panels), as well as the insulation. Since the prefabricated parts make up the entire structure, the total details can serve several purposes if they are complete to the point of showing the wall and roof materials.

SUMMARY

The long-span, beam-and-column, and wall-bearing structural types of steel provide excellent buildings for commercial activities. Fastening of members for such structures may be by riveting, bolting, or welding, or by a combination of the types. For ease and economy of shipping and erection, large units may be partially assembled in the fabricating shop, with the balance of the assembly performed in the field. For this reason, it is important for the detailer to understand the symbols used to indicate the various fastening devices and

to know the shop and field symbols for each.

The confusion created by incomplete or erroneous drawings and details can hamper efforts to estimate, fabricate, ship, and erect structural steel. It is imperative that the detailer understand and use the standard drawing practices and symbols in an effort to communicate fully the necessary information to all those involved in creating a commercial building.

Problems—Chapter 3

1. The bracket shown is to be fabricated using fillet welds. Prepare the details and shop drawings for it using standard welding symbols.

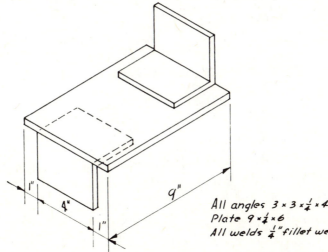

All angles 3 × 3 × ¼ × 4
Plate 9 × ¼ × 6
All welds ¼" fillet welds × 2½"

2. A roof beam is shown with end connection angles attached with shop-applied ⅝-in. rivets. Draw details and shop drawings, and prepare a bill of materials for the beam and connections.

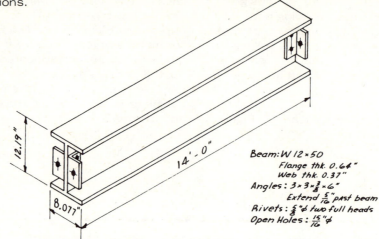

Beam: W 12 × 50
    Flange thk. 0.64"
    Web thk. 0.37"
Angles: 3 × 3 × ⅜ × 6"
    Extend ⁵⁄₁₆" past beam
Rivets: ⅝"⌀ two full heads
Open Holes: ¹⁵⁄₁₆"⌀

3. The truss in the sketch is to be shop-fabricated from steel angles back to back. All joints are to be fastened using ½-in. bolts. Prepare all drawings necessary for the fabrication of the truss.

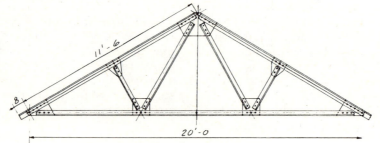

Top Chords: 2 ∟ 3½ × 2½ × ¼"
Bottom Chord: 2 ∟ 2½ × 2½ × ¼"
Web Members: 2 ∟ 2 × 2 × ¼"
Bolts: All joints shop-bolted with ½"⌀ bolts
Bolt spacing: 3" o.c.
Gusset Plates: All plate ⁵⁄₁₆" thk. - size as needed

4. Prepare a complete bill of materials for a single truss as detailed for problem 3.

5. The column shown supports the ends of the two beams attached to it. All fasteners are ⅝-in. bolts applied during erection. The connections are of the framed type. Draw details and shop drawings for the column and connecting angles. Do not include the cap and base plates for the column.

Column: W 8 × 31 × 9'- 0¾ in length
    Flange 0.43" thk, Web 0.29" thk.
Beams: W 8 × 17
    Flange 0.31" thk., Web 0.23" thk.
Angles: 3 × 3 × ⅜ × 5"

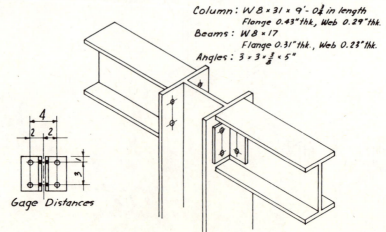

Gage Distances

6. Draw the truss joint detail shown as it would appear for a welded truss using ¼-in. fillet welds each 3 in. long.

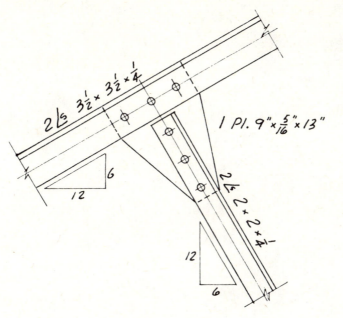

INTRODUCTION

Steel continues to be, as it has been in the past, a logical choice for the structural systems of many industrial buildings. Earlier chapters have explored the virtues of steel and many of the attributes which make it a satisfactory structural material.

The large, undivided space frequently required by industry can be provided by built-up steel units (girders, trusses, and rigid frames), and these components can span large distances with excellent member weight/strength ratios.

Industry moves quickly, and any shutdown time spent waiting on completion of construction is costly. Steel structural systems can be rapidly erected, a factor important in the industrial building field. Industry needs structures which permit flexible use of space in order to facilitate

chapter 4
steel structural systems-industrial

changes in methods, processes, and areas required for its various operations. Buildings based upon steel girder or truss systems can be suitably altered to meet the changing demands for space and its utilization.

Built-up composite steel members, namely girders and trusses, have attained a predominant place in industrial construction. These members are capable of spanning reasonably long distances. When areas are required which are larger than single units can provide, it is not uncommon to run girders or trusses in continuous fashion over several supporting columns. Generally the ends of girders and trusses of industrial buildings are supported by steel columns rather than on load-bearing walls. This practice permits more rapid erection and makes it much easier to alter walls (windows, doors, etc.) at a later date.

COLUMNS IN INDUSTRIAL STRUCTURES

The columns supporting the ends of girders and trusses provide end-bearing for the units and support the wall construction to overcome wind and impact loads. The same units may be called upon to support crane girders, which in turn support the actual track (or rails) of traveling overhead cranes. The columns may be one piece or composite (or stepped) as shown in Figure 4-1.

Figure 4-1 illustrates a third type of column configuration wherein two separate column

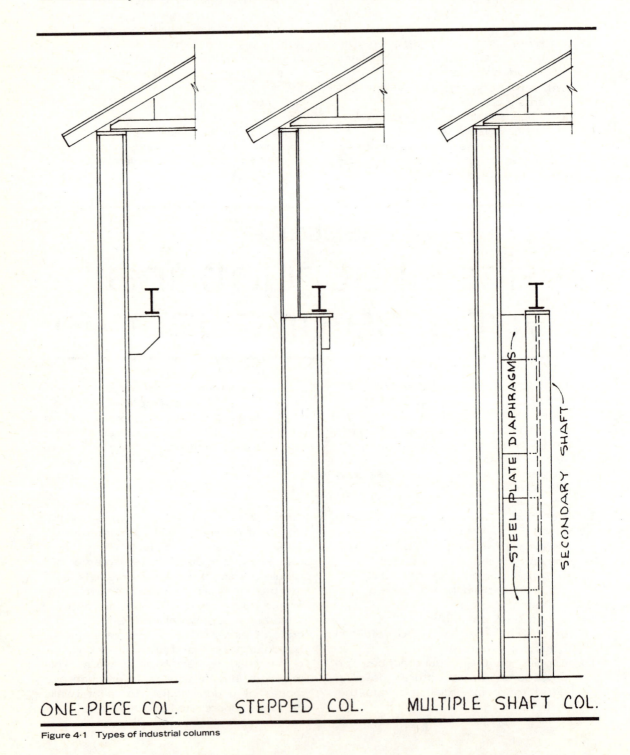

ONE-PIECE COL.   STEPPED COL.   MULTIPLE SHAFT COL.

Figure 4·1 Types of industrial columns

shafts are tied together with steel diaphragms. One shaft supports the roof structure, while the second member supports the crane girders.

Some industrial buildings require side-by-side areas, both of which are to have traveling cranes. The central columns (interior) in such instances may have additional shafts attached to both sides. These two secondary columns provide support for the crane rails on each side in the same fashion shown for an exterior column in the illustration.

THE BENT

The main structural members in a transverse structural frame (the columns and girder, or truss) collectively form what is referred to as a *bent*. Included in a typical bent are the columns and roof members, and the diagonal braces between columns and roof units. These diagonal braces are termed *knee braces* and are primarily intended to resist stresses created by wind loads.

Figure 4-2 shows a typical *girder bent* and a *truss bent*, each with and without knee braces. Notice that the knee braces run from the columns to the bottom chord of the truss or to the underside of the girder.

INDUSTRIAL TRUSSES

A number of truss types are used in industrial construction. The main truss types are

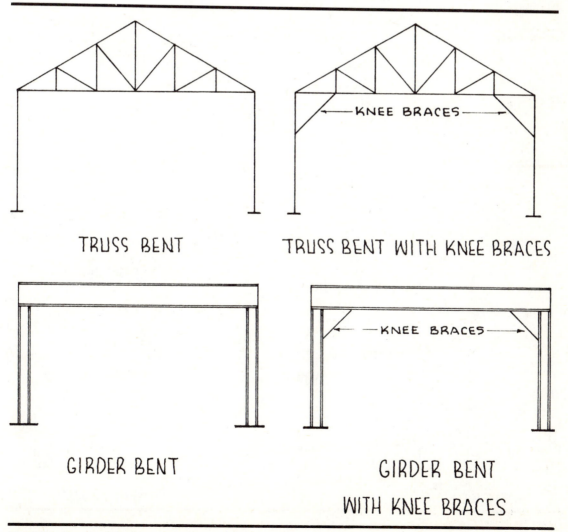

Figure 4·2 Truss and girder structural bents

shown in an industrial configuration in Figure 4-3. The types include the Pratt, Warren, Fink, scissors, bowstring, sawtooth, and Vierendeel trusses. Each type offers characteristics which fit certain applications better than others.

The Vierendeel truss is a totally rectangular unit. It is extremely useful in instances requiring clean joints without intruding bracing members. In some multistory buildings the Vierendeel truss permits free access through the truss itself for mechanical systems and service.

The designer chooses the type of truss to be used based upon a number of considerations. The type of roof desired, whether flat, sloping, or gabled, is a definite factor. Interior space requirements enter into the decision, and the forms of live and dead load acting upon the structure must be considered.

The combination of several types of trusses, girders, or portions thereof in one structure is not uncommon in industrial construction. The sketches in Figure 4-4 indicate a few of the many configurations which can be utilized.

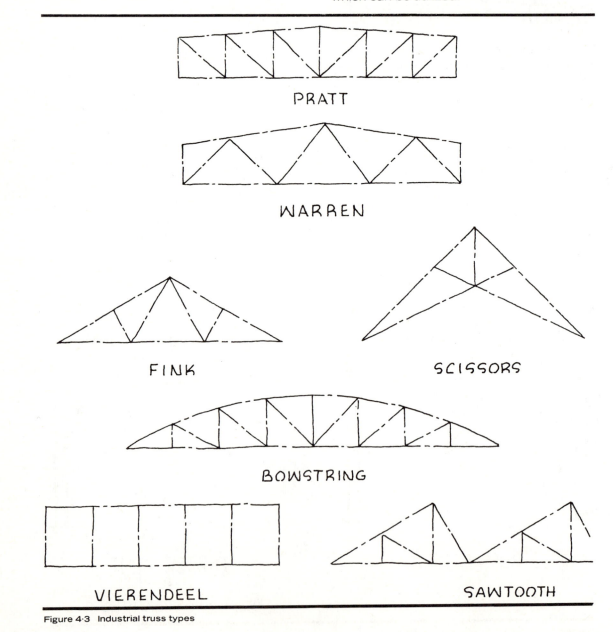

PRATT

WARREN

FINK

SCISSORS

BOWSTRING

VIERENDEEL

SAWTOOTH

Figure 4·3 Industrial truss types

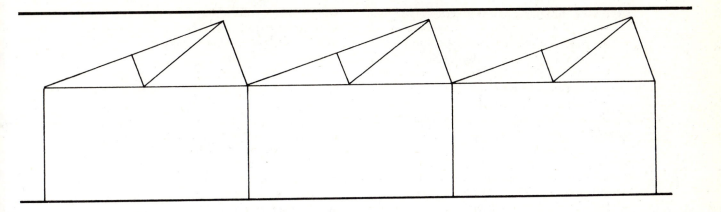

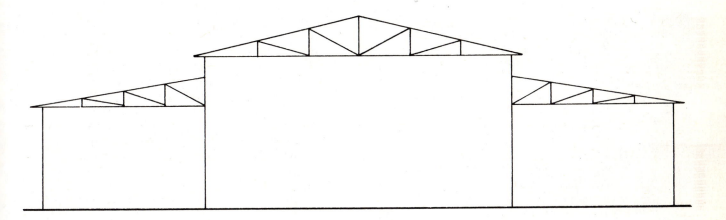

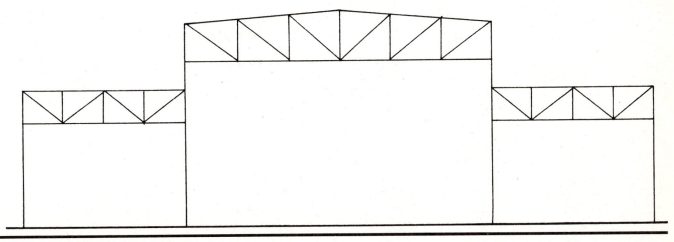

Figure 4·4   Example of industrial building types

LONGITUDINAL MEMBERS

Structural bents of industrial buildings are tied together by the longitudinal and bracing members necessary to make the structure rigid. These members also support wall and roof materials and absorb and transmit the various types of loads (snow load, wind load, etc.).

The sides of a structure are often framed with *girts* (horizontal members attached to the columns) as shown in Figure 4-5. The girts are frequently channels or angles, and the siding material is attached directly to them. The girt spacing depends upon the span between columns and the attachment spacing necessary for the wall panels or siding material. Girts are provided on the endwalls of structures, and it is often necessary to install intermediate columns in the endwall framing. This practice reduces the unsupported lengths necessary for the girts. The intermediate (or supplementary) columns may not carry any of the roof load, but they do support girts, wall panels, and some types of diagonal bracing.

The *purlins* of a structure support the roof in the same manner that girts support the walls. Purlins are longitudinal members which tie the bents together in the roof plane while providing the support for the actual roofing materials or roof panels (see Figure 4-5). Purlins, like girts, are often made from channel or angle shapes. They may run between or across the top chords of trusses (or the tops of the main girders). The purlins installed at the ridge of gabled-roof structures are referred to as *ridge struts*. The purlin units are placed back to back at the ridge and are tied together with diaphragms (steel plates at spaced intervals) or threaded rods so as to form one single structural unit. Like regular purlins, ridge struts support the roofing material and provide additional longitudinal bracing between the bents.

Another longitudinal member similar to the purlins and ridge struts is the *eave strut.* This member is attached to the columns of the bents at the point where the top chord of the truss and the column meet, actually at the eave of the building. Made of channel or angle shape, the eave strut supports both wall and roof materials, as well as providing additional longitudinal bracing. Figure 4-5 illustrates both ridge struts and eave struts.

BRACING

Industrial structures are subjected to relatively heavy wind loads because of their size. Wind loads tend to distort and rack the structural system. Adequate wind bracing becomes an important part of the structural design.

Industrial buildings are subject to shocks and vibrations created by moving equipment of various types, and by traveling overhead cranes in particular. The structural system must have sufficient continuity to withstand the stresses created by such conditions. Bracing is necessary to negate the possibility of damage from such stresses and the wind loadings. Some of the various types of bracing appear in Figure 4-5. Certainly the purlins, girts, eave struts, and ridge struts provide considerable assistance in tying the structure together, but they do not preclude the need for diagonal bracing members.

A standard practice in industrial buildings with traveling overhead cranes is to install a horizontal bracing system between bents in the same plane as the bottom chords of the trusses. Diagonal bracing may be installed in the same location also. A system of diagonal bracing between trusses may also be provided in the same plane as the top chords of the trusses.

Additional bracing is sometimes necessary at the roof in the form of *sway frames*. These braces run in a vertical plane at the center of the trusses and brace between the trusses. The function of sway frames is much like that of the cross-bridging installed between the wood floor joists of residential construction.

Diagonal bracing in sidewalls and endwalls counteracts wind loads and helps to dampen the vibrations set up within the structure. Steel rods or cables can be used for diagonal bracing in steel industrial structures, but they have a tendency to

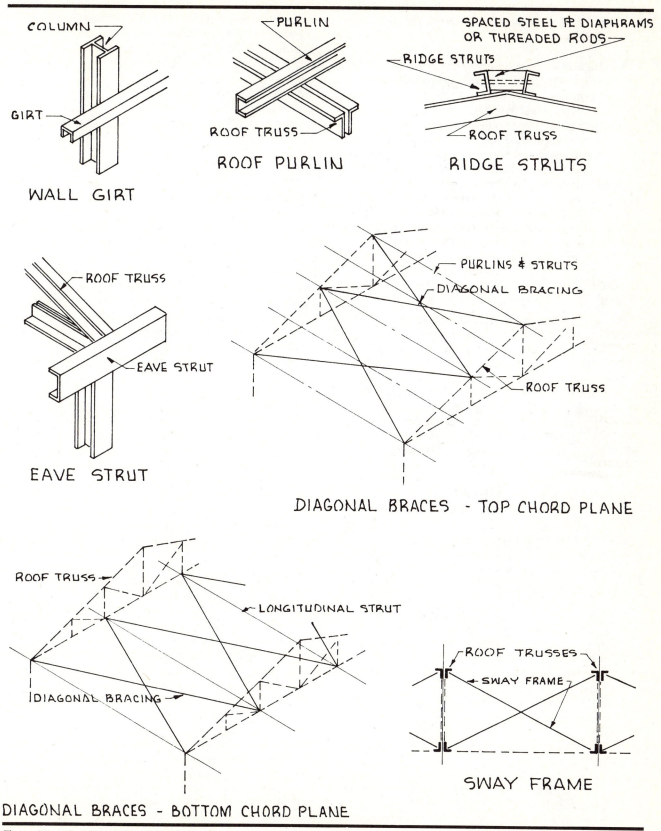

Figure 4·5 Industrial building components

vibrate. As a result, channel and angle sections are often preferred for bracing purposes.

The actual location of diagonal wall bracing varies with the structural system used. It is not generally necessary to brace every bay in this fashion. The ends of the structure should definitely receive diagonal braces, as should the end bays of the sidewall framing. Intermediate bays should be diagonally braced as required for the specific building. In many instances diagonal bracing is provided for the center bay (or bays) of long structures. Diagonal bracing of the various types can serve the additional purpose of providing support for components during the erection process. The braces can hold elements in a true position until the total assembly is complete.

The fastening systems employed in attaching bracing members to the main structural frame are varied. Braces may be fastened by welds, rivets, or bolts. The brace may be attached directly to the other structural members, or it may be attached to a *gusset plate*. The gusset plate is a steel plate provided on the main member, with provisions made for the attachment of the brace to the gusset during erection.

EXAMPLE OF A SIMPLE INDUSTRIAL BUILDING

The simple industrial structure shown in Figures 4-6 and 4-7 is based upon steel bents composed of one-piece wide-flange columns, roof trusses, and knee braces.

Figure 4-6 is a composite plan showing (in the upper half) the framing plan in the plane of the bottom chord and (in the lower half) the plan in the plane of the top chord of the trusses. In addition the figure includes a side elevation clearly showing the sidewall girts and bracing. Notice that since the building is symmetrical about a center line, only the left half has been drawn in plan and elevation.

Sag rods as called out in Figure 4-6 are steel rods threaded at each end to receive structural steel nuts. The rods provide intermediate support (actually eliminate sagging of members) for roof purlins and

sidewall girts. The sag rods in the example are located midway between bents. Larger bay spacing could require that sag rods be provided at more frequent intervals, such as one-third or one-fourth points.

The detail of the column, girts, and end connection for the truss is shown in Figure 4-7. Also included is the design drawing for a complete bent including column, truss, and knee brace. The drawing is in the form of a cross section and includes purlins and girts. Notice that only half the truss appears because of the symmetrical nature of the bent.

The drawings make use of the *d.o.* or *do* (ditto) marking to indicate the similarity of repeating structural components. The cross section carries a typical note stating that all truss members are to be the same section size except where noted. The use of such a note is time-saving. It also spells out the information more clearly than would repeated individual labeling of each similar section.

General notes as included on Figure 4-7 are used to convey a variety of information. The referral source for specifications, for material designations, and for fastening methods and devices is often included in the general note category.

EXAMPLE OF AN INDUSTRIAL BUILDING WITH CRANE

The building represented in Figures 4-8 and 4-9 is another typical example of an industrial structure based upon steel structural members. This particular structure includes provisions for an overhead traveling crane. Since the crane girders must be supported, as well as the roof structure, the columns in this instance are of the stepped column type. The roof trusses are of the Pratt type and are fabricated from standard steel components.

The plan and elevation indicated in Figure 4-8 provide the design drawings for the structure, along with the cross section and partial column detail in Figure 4-9. These drawings follow a format similar to that used in the previous example.

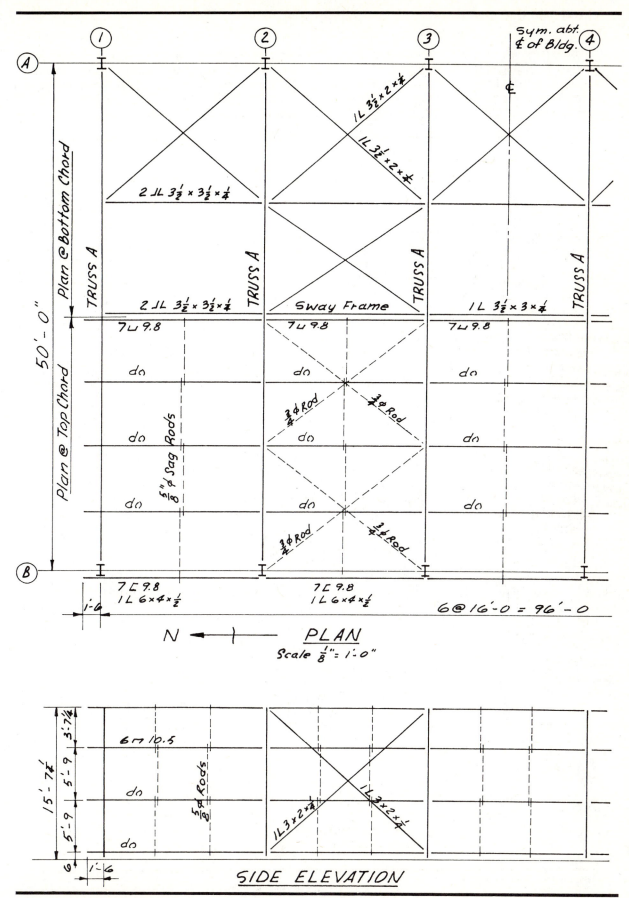

Figure 4·6   Industrial building details

The extremely low roof slope precludes the need for sag rods between the roof purlins Sag rods are, however, still a necessity in the wall system and are employed to provide support between the horizontal wall girts. A single member can be installed at the ridge in lieu of the double-member roof strut of the previous building. This is possible in this case because of the low roof slope.

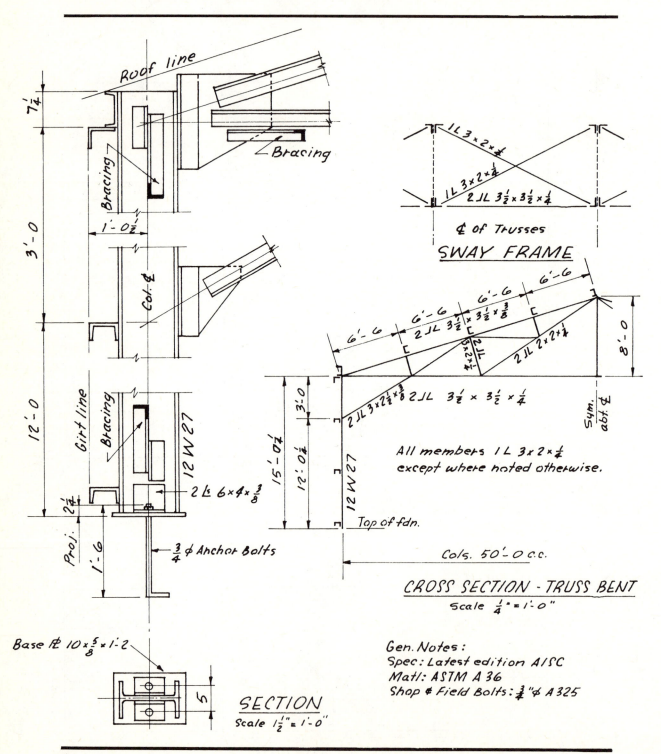

Figure 4·7 Truss details

Diagonal bracing in the plane of the bottom chords of the trusses, and sway frames between trusses, combine to provide the bracing necessary to resist stresses created by the movement of the crane and the wind load on the exterior of the building.

Bracing between the columns of the bents does not, in this building, rely upon the

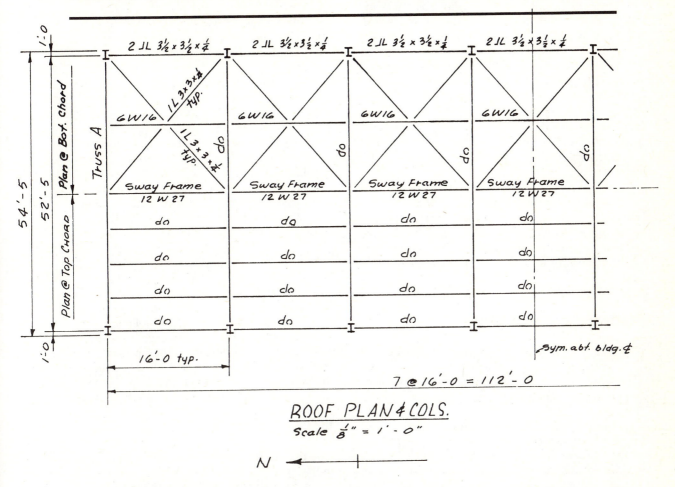

ROOF PLAN & COLS.

Scale ⅛″ = 1′-0″

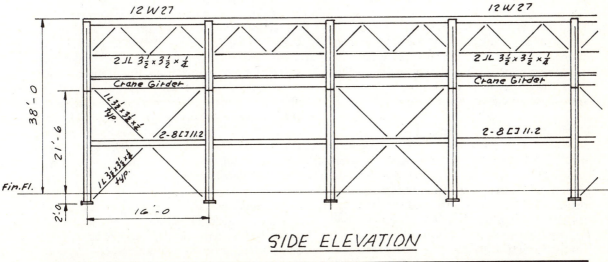

SIDE ELEVATION

Figure 4·8 Sample industrial building details

sidewall girts alone. Heavy steel channels, facing each other, are provided between the columns to form horizontal members for the purpose of resisting the stresses created by the crane. The two channels are joined at intervals by steel bar stock installed at the top and bottom surfaces of the channels.

The actual track (or rail) of the crane is supported by the heavy wide-flange sections which serve as *crane girders*. These members are attached to the columns and bear directly upon the bottom section of the stepped columns. The girders are also attached to the side of the upper section of the columns

In typical fashion, the drawings for this structure carry call-outs and notes. These provide dimensional data, specifications, and information regarding materials, methods, and fasteners.

TYPICAL SHOP DRAWINGS

Once the design drawings have spelled out the intended structural system, shop drawings can be prepared for the various structural components. The trusses for the building shown in Figures 4-8 and 4-9 provide the basis for the partial shop drawing of the truss in Figure 4-10. For the actual structure, shop drawings would be prepared for all structural members in the same manner.

The Pratt truss is drawn in detail in elevation. The separate view of top and bottom chords are included with the elevation. The top chord is drawn as viewed from above, while the bottom chord is shown as it would appear when viewed from below.

Members such as *webs* (diagonal units) may be made up of the same size angles but may vary in length. It is important on shop drawings to indicate the section shape, size, and length of each member, since this information is required for ordering materials and fabricating the units. Notice that on the example the information is given next to each individual member to avoid any confusion as to the stock required.

Fasteners can be combined in various manners. It is much more consistent and clear to rely upon a single size and type of fastener for the entire truss. The different stress values at points throughout the truss can then be met by the number of fasteners used at each joint, rather than by changing the size or type of fastener to meet each individual condition.

The fasteners in the example, and the open holes provided for field erection, are of uniform size and are called out in the notes. It should be clear how much less confusing this practice is than labeling every fastener and hole individually. The critical point in this instance is to be certain that all fasteners and holes have been indicated with the proper symbol and that their spacing and location have been clearly designated during the drawing process.

The truss in Figure 4-10 is symmetrical about the center line, so in the total drawing it is permissible to detail only half of the truss. The initial layout of the drawing is based upon establishing the critical dimensions, the intersecting panel points, and the gage lines recommended in the standards or manual for the individual members. The finished drawing is then developed from the layout drawing, with actual members and fasteners shown, as in the example.

The practice of drawing only half of a symmetrical truss is well established, and most firms require that the left half be drawn. It is again advantageous to conform to the common industrial practices throughout the detailing process to avoid confusion. Most structural workers are used to seeing the left half rather than the right on the drawings. The detailer must be aware of any differences which may exist between the left and right halves of a truss, and the entire truss should be detailed if there are such dissimilarities. When viewing the example, keep in mind that it is only a portion of the total drawing.

Information is sometimes added to the detailed drawings of trusses. The drawings must spell out clearly any degree of camber required in the bottom chord, and the dimensions and drawing should clearly

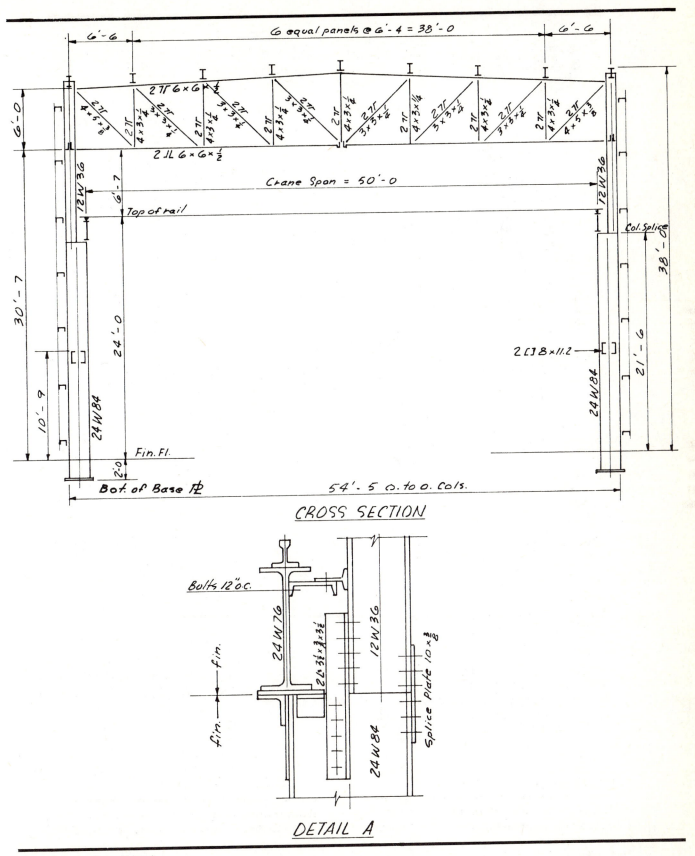

Figure 4·9 Section and detail

After Studying This Chapter You Should Be Able To:

1. Describe some of the earliest wood structural forms for residential use.
2. List some of the factors which severely limited early builders in their use of wood.
3. Describe the fibers which are a part of the makeup of wood.
4. Explain the components making up the material called wood, and explain the functions of the components.
5. Name the six factors of composition that can be identified in the cross section of a tree, and explain how they serve the tree.
6. Define *growth rings* (or annual rings).
7. Describe *springwood* and *summerwood* and point out their differences.
8. Explain the general basis used in determining lumber grades.
9. Define the categories of softwoods and hardwoods.

19. Describe the various fastening devices and methods available for wood truss fabrication.
20. Describe plywood box beams and stressed-skin plywood panels.
21. Describe structural glued laminated lumber, and cite its advantages over regular lumber.
22. List some advantages and limitations of wood as a structural material.

INTRODUCTION

For thousands of years, man has used wood as the basis of various structural systems. Early structural forms used wood with grass and mud (wattle) to create enclosures of relatively small size (much as the beaver builds its shelter). The hogans still prevalent on some American Indian reservations and the structures of some tribes in Africa and Australia retain the methods employed by those earlier builders (Figure 5-1). The dwellings of the Plains

chapter 5
the material-wood

10. Describe the process of curing (or drying) wood, explain its function, and describe the effect it has on the material.
11. List and explain the three types of natural infirmities in wood which are avoided, as much as possible, when selecting lumber for structural purposes.
12. Define the terms *close*, *medium*, and *coarse* grain.
13. Describe *cross* grain, *edge* grain, and *flat* grain.
14. Explain how nominal and actual sizes vary in structural lumber.
15. Explain the significance of the S1S, S2S, S3S, and S4S symbols.
16. Describe *engineered* lumber, and explain how it differs from regular dimension lumber.
17. Name the stresses usually given in design tables for various grades and species of lumber.
18. Cite the advantages provided by wood trusses.

Indians were often supported by a pyramidal frame of straight tree branches, or saplings, laced together with rawhide and covered with animal skins (Figure 5-2).

Straight logs (*vigas*) were used extensively by the builders of the Southwest in their adobe structures as the main roof framing system, with small logs fitted together to span them and form the roof deck (Figure 5-3). We find architects today designing similar structures in the Desert Southwest.

The log structures erected by settlers, trappers, and miners in the days of the early West relied on load-bearing walls of horizontal logs. The roof structures employed full (or half) logs as rafters or roof joists (Figure 5-4). The bridges of that era were often of post-and-beam concept, with logs serving as both types of structural members. The largest limitation on these early builders was that imposed by the limited spanning capabilities and the immense weight of the log members.

Figure 5·1   Structures using "wattle" construction

Figure 5·2   Plains Indian shelter

Progressing from rough logs to milled lumber, wood has continued through the ages to serve man as a versatile and practical structural material for his structures. The use of wood structural systems is widespread today, and the sophisticated technology of wood construction has enabled today's designers to accomplish things with wood undreamed of by earlier man.

COMPOSITION OF WOOD

Wood, unlike many other structural materials, is made up of hollow tubes or cells (many times longer than their width). These cells are called *fibers*. In dry form, as with the lumber suited for construction purposes, the cells are mostly empty and hollow. Fiber length varies from about $1/25$ in. for hardwoods, to as much as $1/3$ in. for softwoods. The composition of wood is about 60 percent cellulose, 28 percent lignin, and

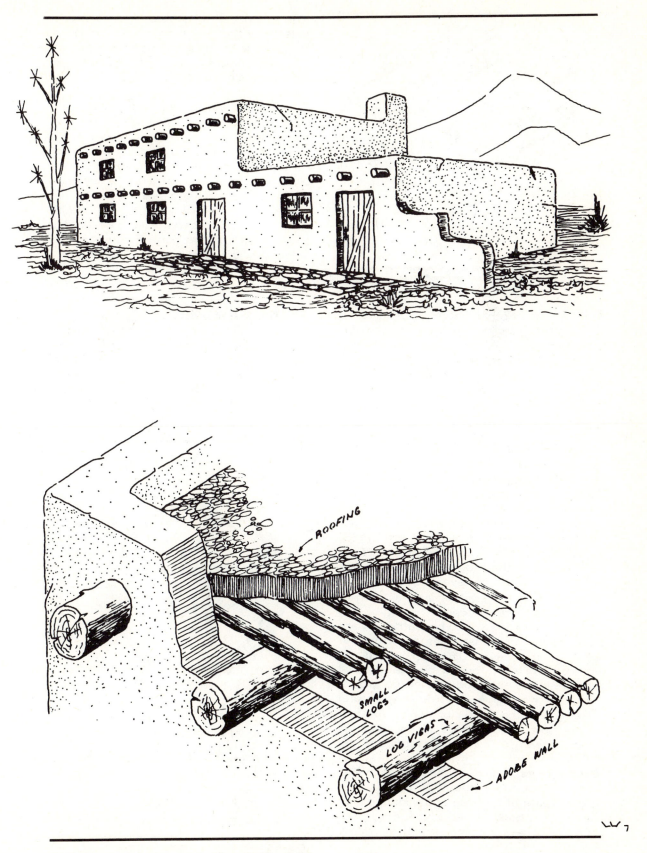

Figure 5·3   Abode structure and roof construction detail

Figure 5·4 Log cabin corner and roof detail

the balance of other materials. Cellulose is the material forming the framework of the fiber wall, and lignin acts as a cement binding the fibers together. The strength, interestingly enough, is dependent upon the thickness of the fiber walls, rather than upon the fiber length.

FORMATION OF WOODS

Figure 5-5 shows a tree trunk in cross section, with its various components labeled. The *outer bark* (A) is of course the corklike layer of dry, dead tissue protecting the tree from external injury. The *inner bark* (B) is moist and soft and carries the food from the leaves to the growing portions of the tree. A microscopic layer just inside the inner bark is the *cambian layer* (C), which forms wood and bark cells. The *sapwood* (D) is the light-colored wood beneath the bark responsible for transporting sap from the roots to the leaves. The *heartwood* (E) is an inactive layer formed by the gradual change in the sapwood as the tree grows. The heartwood gives the tree its structural strength. In the center of the tree is *pith* (F), a soft tissue about which the first wood growth occurs in newly formed twigs.

Growth rings Notice on the section in Figure 5-5 that concentric rings start at the center (pith) of the tree and continue outward to the bark. For each year of growth of a tree, a new ring (annual ring) is formed.

The growth is originated in the cambian layer, and the year's growth is formed on the outside of the previous year's growth. The bark, naturally, is pushed outward by each year's growth.

Springwood and summerwood Many species of wood have each annual ring divided into two layers; the inner layer called *springwood* consists of fibers with large cavities and thin walls, whereas the outer layer known as *summerwood* is generally heavier, harder, and stronger. It has smaller fibers with thicker walls. Specific gravity, or wood density, is measured in terms of the proportion of summerwood. Generally the higher the density, the greater the strength of the wood.

LUMBER GRADES

Growth rings and the proportion of summerwood to springwood are often used as a visual method of selecting lumber of different grades. Various societies and associations have stringent rules affecting the

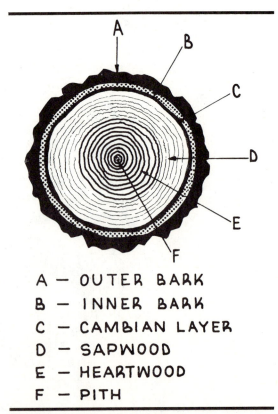

A — OUTER BARK
B — INNER BARK
C — CAMBIAN LAYER
D — SAPWOOD
E — HEARTWOOD
F — PITH

Figure 5·5 Tree trunk in section

grading process. Typically one might find a definition of *dense* for a certain species as "having no less than six annual rings per inch and a minimum of one-third summerwood measured over a 3-in. portion of a radial line." Since strength bears a close relationship to the density of the wood, it follows that lumber of exceptionally light weight is generally excluded from structural applications.

HARDWOODS AND SOFTWOODS

Trees fall into two classes: the hardwoods and the softwoods. These classifications may be initially misleading, since many hardwoods are softer than some softwoods, and some softwoods are as hard as the average hardwood. A clearer definition lists the softwood tree as "one of the botanical group of trees that in most cases have needle or scalelike leaves" (also termed *conifers* because most softwoods bear cones). The hardwood tree, then, is set apart by being in the botanical group of trees generally having broad leaves. The term *deciduous* often applied to hardwood trees simply refers to the fact that the trees drop their leaves during certain seasons of the year. The cedar, pine, redwood, spruce, fir, hemlock, larch, and cypress species are classed as softwoods, with the balance of tree species falling in the hardwood category.

CURING OF WOOD

In its natural state in the living tree, wood contains a large quantity of water, or sap. When converted to lumber, it begins to lose the moisture. Lumber mills employ both air-drying and kiln-drying to reduce the moisture from the cavities and the fiber walls. This drying process is necessary for stability and is a very effective method of increasing the strength of the wood. An average air-dried piece of wood is as much as 2½ times as strong as it was in the green, or just cut, condition. When moisture content is reduced, the wood shrinks, and the shrinkage is most in the direction of the growth rings, about one-half as much across the rings, and very little along the grain.

IMPERFECTIONS

In selecting lumber for structural purposes, three types of infirmities in the wood are generally avoided if possible. *Knots* (that portion of a branch that has become incorporated in the tree body) influence the strength of lumber. In selecting lumber the size and location of knots are severely limited when the particular piece is to be subjected to extreme stresses. *Checks* (see Figure 5-6) are lengthwise separations of wood, the largest part occurring across the growth rings. *Shakes* (Figure 5-6) are separations along (or with) the grain and run between the growth rings. Both checks and shakes reduce the strength of lumber where bending may occur (as with floor joists, beams, etc.) but have little effect on the strength of members used for longitudinal compression.

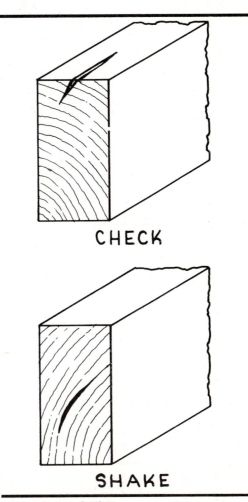

CHECK

SHAKE

Figure 5·6 Imperfections

GRAIN

The very loosely applied term *grain* used in describing wood products can be best understood by studying the various types of grain. The terms *close grain*, *medium grain*, and *coarse grain* are used to describe the texture of the wood. They simply refer to the width and spacing of the annual growth rings in a given piece of lumber.

Cross grain, *edge grain*, and *flat grain* are illustrated in Figure 5-7. Notice that cross

grain occurs when the direction of the wood fibers is not parallel to the edges of the piece of lumber. Edge grain is the term applied to lumber in which the growth rings form an angle of 45° or more with the face of the piece, and flat grain is applied to the piece in which the growth rings are approximately tangent to the face (or form an angle of less than 45° with the face).

SIZE AND FINISH

Most structural lumber is called out and ordered by its nominal size. This size indicates the actual dimensions of the piece before milling (or surfacing) the various sides for smoothness and appearance. In most cases, then, a 2×4 is actually $1\frac{5}{8} \times 3\frac{5}{8}$ in. when it arrives at the construction site. The decrease in size from nominal to actual is generally $\frac{3}{8}$ in. (and sometimes $\frac{1}{2}$ in.) in each dimension. Lumber may be purchased with surfacing done on only part of the sides. The most common structural lumber is surfaced on all four sides, symbolized by the S4S designation, but lumber can be specified S1S, S2S, and S3S as well.

A relatively new concept in the lumber industry is that of providing *engineered* lumber. Through a process of selecting extremely high strength lumber and curing it to exacting specifications, this lumber provides the strength of larger pieces while running consistently smaller in size. Engineered 2×4s could thus be $1\frac{1}{4} \times 3\frac{1}{4}$ in. or $1\frac{1}{2} \times 3\frac{1}{2}$ in. rather than the usual $1\frac{5}{8} \times 3\frac{5}{8}$ in. stock.

Regular structural lumber is classified by size into the following categories:

Beams and stringers: Lumber with rectangular cross section, 8 in. or more in width and 5 in. or more in thickness.
Joists and planks: Lumber with rectangular cross section, 4 in. or more in width and 2 to 4 in. in thickness.
Posts and timbers: Lumber with approximately square cross section, 5×5 in. or larger.

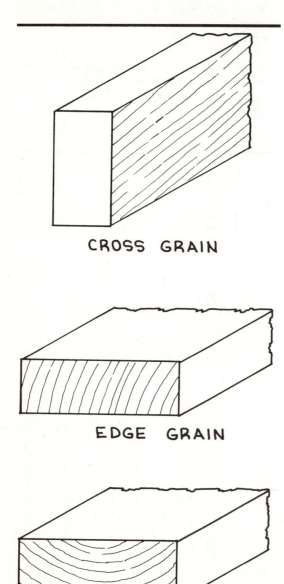

CROSS GRAIN

EDGE GRAIN

FLAT GRAIN

Figure 5·7 Grain types

TABLE 5-1 ALLOWABLE UNIT STRESS (PSI) FOR DOUGLAS FIR (DFCR)*

SPECIES	Use	Grade	f and t	H	c	c⊥	E
D.F.C.R.	Light Framing	Dense Select Structural	2,050	120	455	1,500	1,760,000
		Select Structural	1,900	120	415	1,400	
		1,500 f Industrial	1,500	120	390	1,200	
		1,200 f Industrial	1,200	95	390	1,000	
	Joists and Planks	Dense Select Structural	2,050	120	455	1,650	
		Select Structural	1,900	120	415	1,500	
		Dense Construction	1,750	120	455	1,400	
		Construction	1,500	120	390	1,000	
	Beams and Stringers	Dense Select Structural	2,050	120	455	1,500	
		Select Structural	1,900	120	415	1,400	
		Dense Construction	1,750	120	455	1,200	
		Construction	1,500	120	390	1,000	
	Posts and Timbers	Dense Select Structural	1,900	120	455	1,650	
		Select Structural	1,750	120	415	1,500	
		Dense Construction	1,500	120	455	1,400	
		Construction	1,200	120	390	1,200	
	Decking	Select Dex	1,500		390		1,760,000
		Commercial Dex	1,200		390		

* Compiled from FHA and industrial information.

STRENGTH

Strength values for various grades and species of lumber are available in tabular form from a number of engineering and lumber industry associations. Since the actual design for stresses is best left to the engineer, this text will not become involved with the detailed design process. The technician should be aware, however, of the terminology used to describe the various values of strength. Generally the tables list the grades available for each species and the allowable unit stresses in psi for each grade. Unit stresses are listed for:

Extreme fiber in bending (f)
Tension parallel to grain (t)
Horizontal shear (H)
Compression perpendicular to grain ($C\perp$)
Compression parallel to grain (c)
Modulus of elasticity (E)

A typical table of allowable stresses would list the various species and all grades for each. In this case Table 5-1 represents that part devoted to the single species of douglas fir, coast region (DFCR).

Knowing the loads involved and the type of stress placed on the member, the structural designer finds the allowable unit stresses from the table and can then choose the species, grade, and size for the particular piece. The diagrams in Figure 5-8 show how the various types of stress act upon the structural member.

WOOD TRUSSES

The natural limitations of size of lumber, and its relatively low load-carrying ability when compared with other structural materials such as steel, led designers to combine numbers of lumber pieces into *trusses*. The truss allows much greater spans, while keeping the weight of the spanning component lower (through the use of smaller pieces fastened together). The efficiency of the truss itself, plus the economics of in-shop construction of the components and fast job-site erection, have made truss use widespread.

Many truss forms are produced, using standard sizes of structural lumber. You will recall the truss illustrations in the previous unit on steel. Wood trusses are produced in the same configurations, including Fink, Howe, Pratt, scissors, bowstring, Warren, sawtooth, shed, and factory types.

Wood truss fastening devices are many and varied. They range from the nailed, glued and nailed, and simple bolted types to the more sophisticated shear plates,

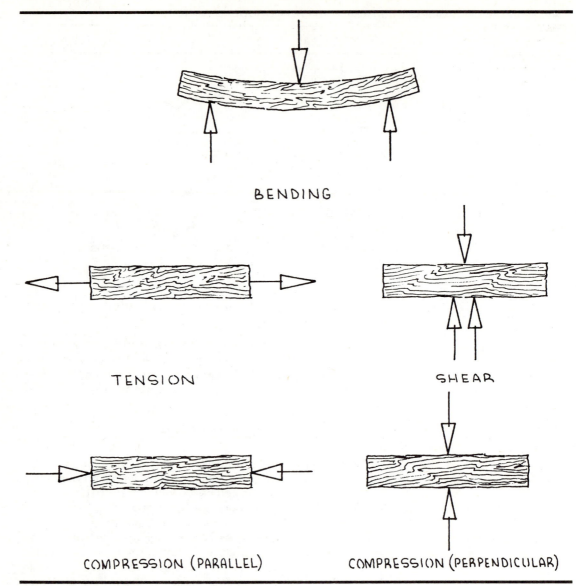

BENDING

TENSION

SHEAR

COMPRESSION (PARALLEL)

COMPRESSION (PERPENDICULAR)

Figure 5·8 Simple stresses

rings, and spiking or nailing grids (Figure 5-9). The sole purpose of the various types of fasteners is to distribute the load at the joint in such a way that the concentration of stress at that point is within the tolerance limits of the wood itself.

BOX BEAMS

Relatively long distances may be spanned by wood members termed *plywood box beams* (Figure 5-10), which combine structural lumber and plywood. These beams are lightweight, are generally shop-fabricated,

and lend themselves readily to tapered, arched, and curved forms. These components span distances of 20 to 80 ft easily, when properly designed for that purpose.

STRESSED-SKIN PANELS

Similar in design concept to the plywood box beam is the stressed-skin panel, made up of structural lumber ribs (or joists) covered on both sides with plywood which is glued and nailed to them. These highly versatile building components are used for floor, roof, and wall construction. In many

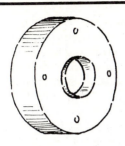

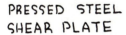

PRESSED STEEL
SHEAR PLATE

MALLEABLE
SHEAR PLATE

SPLIT RING
CONNECTOR

CLAW PLATE
WITH HUB

CLAW PLATE
WITHOUT HUB

SPIKE GRID

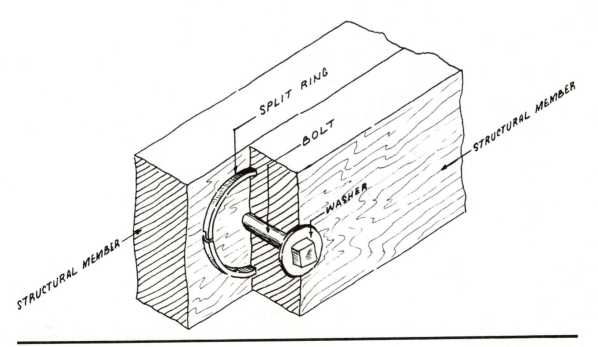

Figure 5·9 Fastening devices for wood

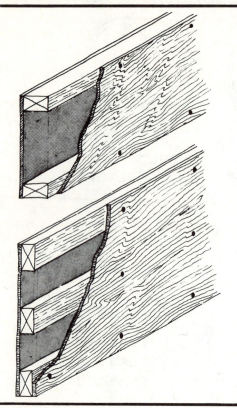

Figure 5·10 Plywood box beams

cases the panels are fabricated with insulation between the plywood skins (Figure 5-11). These shop-fabricated units provide a means of extremely fast on-site erection, a factor greatly appreciated in times of inclement weather. Such panels (with 2 × 2 joists) can serve in residential applications as floor structure spanning as much as 28 ft and roof structure spanning up to 32 ft, depending on load conditions. To develop the necessary stress potentials, great care is necessary in the glue section and in the nailing and gluing process.

LAMINATED WOOD

A most effective structural component has been developed by gluing (laminating) individual pieces of structural lumber together. Termed *structural glued laminated lumber*, these sections provide a much more homogeneous material and greatly increase the capabilities of heavy timber construction by creating more uniform physical characteristics throughout the piece. Many of the structural drawings are still prepared by the engineering or architectural firm, but the

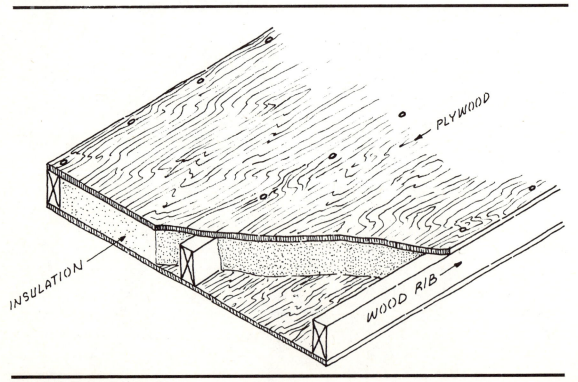

Figure 5·11 Stressed skin panel

bulk of the structural design calculations of laminated wood construction is done by the manufacturer's engineering staff.

The most common lamination system is that of combining pieces in which the grain all runs approximately parallel and in the longitudinal direction. Other combinations used frequently are illustrated in Figure 5-12. Many laminated wood items are available as stock (or standard) beams often referred to as *billets*. It becomes a custom item production for arch sections, rigid frames, and special beams, since these items are usually designed and created for a specific combination of load requirements and spans. Naturally enough, the greatest economy in laminated lumber occurs when the piece is made up of multiples of standard dimension lumber.

The most basic forms of laminated material are shown in Figure 5-13 and include the straight member, the rigid frame, the radial arch, and the three-hinged arch. Laminated wood decking is used for floors and roofs and has the advantage over solid timber decking in that the exposed lamination can be specified for a better appearance (as for the finished ceiling).

LIMITATIONS

Wood can be a satisfactory structural system provided that it lends itself to the type of installation proposed and that its disadvantages have been assessed and determined not to be detrimental in the particular case. Wood is naturally combustible, and wherever this factor is of prime impor-

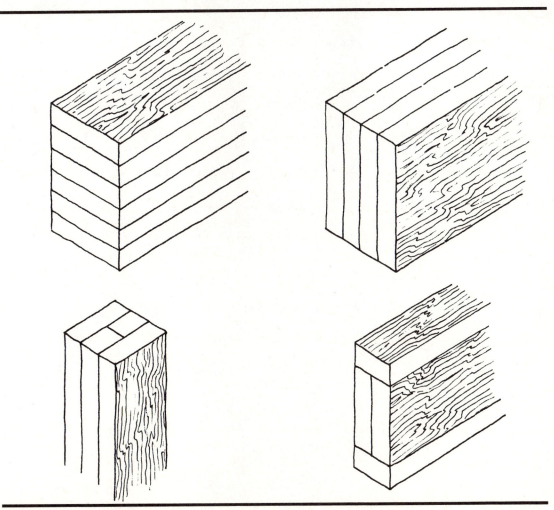

Figure 5·12 Wood-lamination systems

STRAIGHT BEAM

PITCHED BEAM

DOUBLE-TAPERED PITCHED BEAM

RADIAL ARCH

ROOF DECKING

RIGID FRAME

THREE-HINGED ARCH WITH OVERHANG

THREE-HINGED ARCH

Figure 5·13 Laminated wood products

tance either treatment for flame retardance should be used or another material should be specified. Laminated and heavy timber construction are much less prone to fire damage and carry a reasonable fire insurance rating.

Wood should not be used in certain areas without protecting it from decay and/or termites. Liquid treatments and pressure treating help in this regard, and termite shields are a very common building practice in the affected areas of the country.

SUMMARY

Wood as a structural system is quite easily worked with, can be erected quite easily by a majority of construction firms, has the natural beauty desired in much of today's architecture, and is relatively inexpensive when used within its physical capabilities. Its use is applicable to residential, commercial, and industrial structures. Modern structural wood technology has developed, and will continue to develop, new and more sophisticated structural systems. Ranging from simple beams and trusses to the arch, rigid frame, lamella, and geodesic laminated configurations, wood enables today's designers to create clear-span structures with spans of as much as 250 ft.

Problems—Chapter 5

1. Prepare a sectional view (at 1 in. = 1 ft 0 in.) of a 36-in.-diameter tree trunk similar to Figure 5-5. Indicate the outlines of 2 × 6s where they would be cut to have (a) edge grain and (b) flat grain.
2. Prepare a series of five sketches (½ in. = 1 in.) showing a 2 × 8 in rough form and how it changes to become (a) S1S, (b) S2S, (c) S3S, and (d) S4S dimension lumber.
3. Prepare a pictorial sketch of a small lawn storage building utilizing stressed-skin plywood panels.
4. Draw a pictorial sketch of an outdoor picnic shelter (with roof only). Use wood roof decking and laminated wood structural members.
5. Conduct a search of manufacturers' literature and reference sources for laminated wood products, and prepare small freehand sketches of the various shapes of laminated arches available.

After Studying This Chapter You Should Be Able To:

1. Name the two most common residential framing systems, and point out their differences.
2. Describe panelized framing systems.
3. Define *post-and-beam* construction, and cite its advantages.
4. Describe *plank-and-beam* construction, and note its desirable features.
5. List the separate drawings that typically make up a set of residential plans.
6. Describe a framing plan, and list the various information it contains.
7. Define the purpose of details, and tell how to determine when they are necessary.
8. Describe the symbols, abbreviations, and call-outs appearing on floor-framing and roof-framing plans.

standard today were unheard of even a decade ago. Actually, most structural detailing for wood-frame residences is done as an integral part of the architectural working drawings. It is desirable, however, that any technician involved in structural detailing have at least a brief exposure to the structural aspects of residential construction.

In this section, several types of systems will be mentioned; however, the student is advised to keep in mind that a number of other systems, or deviations from those shown, are used in certain locales.

CONVENTIONAL SYSTEMS

Certainly the "old reliable," and probably the most-used, wood structural systems employed are the *western* or *platform* fram-

chapter 6
wood structural systems-residential

9. Define the term *roof pitch*, and give a typical symbol with its interpretation.
10. Describe the drawings necessary to detail a wood truss properly.
11. Describe the procedures followed in detailing a wood truss.
12. Explain how the various components of post-and-beam and plank-and-beam framing systems are shown in plan and detail.
13. List the important factors to consider in detailing a special structural system.

INTRODUCTION

The wood structural systems currently employed in residential construction are numerous and diverse. The technology of structural wood design has progressed rapidly, and many practices considered

ing method (Figure 6-1) and the *balloon* framing method (Figure 6-2). Derivations of these two systems are sometimes called *modern braced* or simply *braced* framing. Attention is directed to the two illustrations, where component terminology and system configuration can be studied. The most significant difference between western and balloon framing is in the system of supporting the floor joists. Balloon framing places the second-floor joists on an inlet 1×6 ledger (ribbon), with most of the support gained by nailing through the joist into the side of the adjacent wall stud. Western or platform framing provides a continuous girt (two 2×4s running over the top of the studs) upon which the second-floor joists rest. For those joists making up the first (or ground) floor, western framing stops the joists short of the sheathing to allow the 2-in. header to run by the joist ends. Balloon framing allows joists to run through to the sheathing with no header involved.

One finds contractors and carpenters creating excellent cases for using one method rather than the other. There appear to be equal numbers extolling the virtues of both, and balloon and western framing appear in almost all areas of the country. Even many of the precut and prefabricated systems do, in essence, incorporate many of the attributes of both methods. The roof framing in the two methods can be of any type from conventional rafters- and-ceiling joists to prefabricated trusses. Most residences having a brick exterior are in reality either balloon or western framed; the brick is applied as a veneer with no structural purpose, primarily for the appearance factor alone.

PANELIZED SYSTEM

A number of *panelized framing systems* have

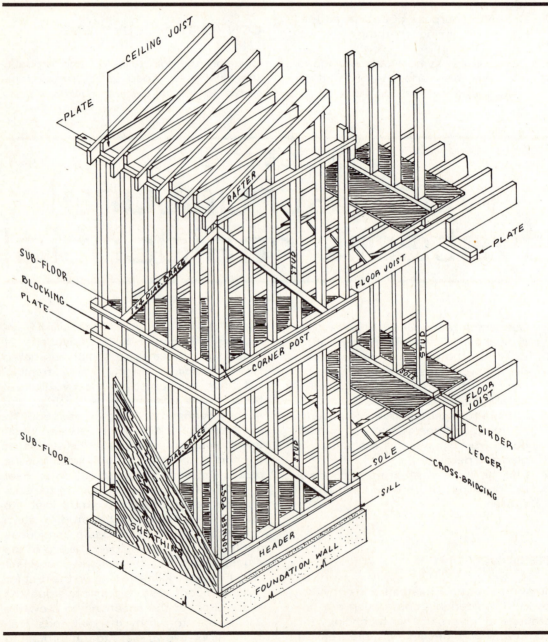

Figure 6·1 Western or platform framing

been developed for residential use, and most rely on modules of 48-in. width, combining in one panel the studs, plates, and sheathing (in most cases plywood). A somewhat typical panel system is shown in Figure 6-3. In this example, the panels are topped by a continuous plate which actually serves as a beam. The double 2 × 4 studs occurring at each panel joint do, in fact, form columns (or posts). The combina-

tion of top plates and columns results in a form of post-and-beam construction (although admittedly impure).

POST-AND-BEAM SYSTEM

The true post-and-beam construction illustrated in Figure 6-4 relieves load-bearing on walls entirely and allows wall construction to serve merely the enclosure function.

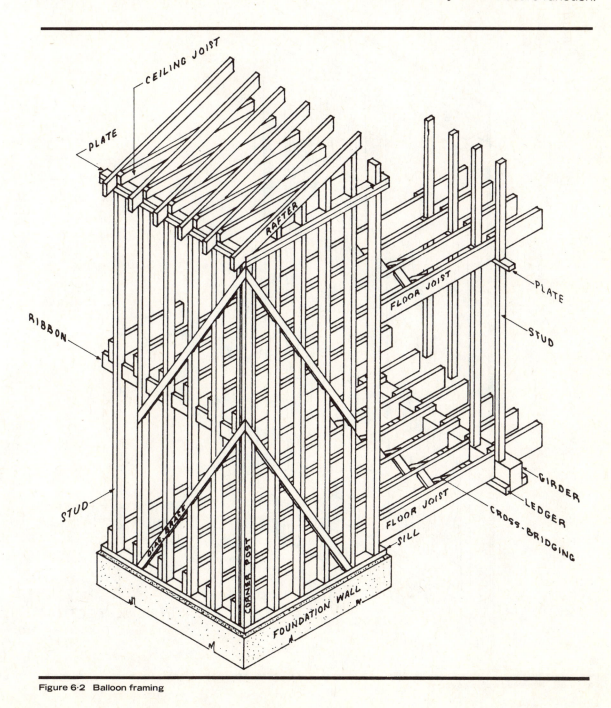

Figure 6·2 Balloon framing

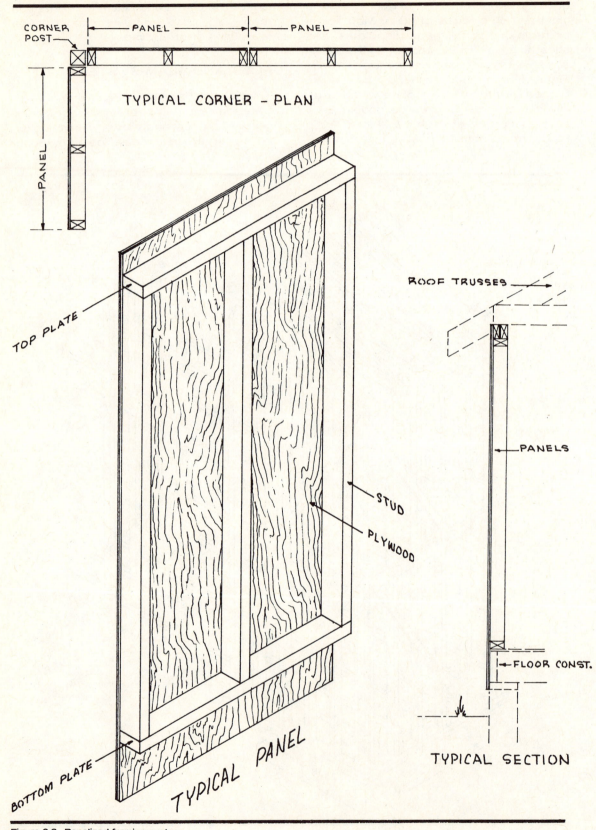

CORNER POST

PANEL

PANEL

PANEL

TYPICAL CORNER - PLAN

TOP PLATE

ROOF TRUSSES

STUD

PLYWOOD

PANELS

BOTTOM PLATE

TYPICAL PANEL

FLOOR CONST.

TYPICAL SECTION

Figure 6·3 Panelized framing system

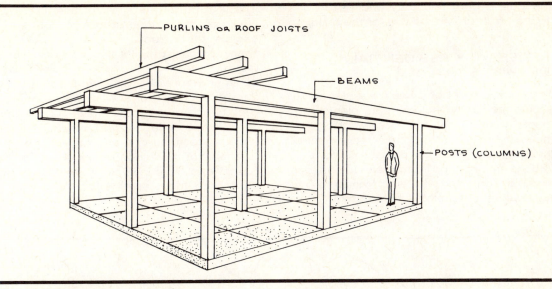

Figure 6·4 Post-and-beam construction

Today's open planning, and use of large expanses of glass, lends itself particularly well to the post-and-beam system. Use of 48-in. modules permits plywood siding, paneling, and gypsum wallboard (Sheetrock) to be used without fitting and cutting, thus taking best advantage of the materials' laborsaving attributes. This type of structural system is quickly erected and, in many cases, saves on the total amount of material necessary to construct the residence.

PLANK-AND-BEAM SYSTEM

Tied very closely to the post-and-beam concept is the system for floor and roof construction called *plank-and-beam* framing (Figure 6-5). The use of laminated wood members is most applicable to the post-and-beam and plank-and-beam methods. Plank-and-beam framing utilizes beams, at reasonably large distances apart, which support a slab (or deck) made up of planks (or decking). The solid timber planks can span considerable distance and are produced with a variety of edge configurations and wood species. Laminated decking is often used for this purpose. The decking tends to form a monolithic unit and reacts to load much as does a poured concrete slab or deck. The resultant structure is most sturdy, with planking (decking)

serving as finished ceiling below and as structural deck (replacing joists and rafters) and sheathing above. The added advantage of the wood deck's natural insulating quality makes this application most inviting to the designer.

Modern trends toward the large roof overhang on residential structures have again shown an advantage for the plank-and-beam system, since the ends of the decking can extend somewhat large distances (cantilever) past the supporting wall beam. The *A-frame* system (Figure 6-6) of building is in many cases an extension of the plank-and-beam method. These very pleasant structures rely on the decking and beams for their entire framing system (except for endwalls). In the case of the A-frame, the sloping roof construction actually becomes both roof and sidewall in many instances.

LAMINATING ADHESIVES

When using laminated members as previously mentioned, it is important to specify that a *weatherproof* glue be used for any members extending outside of the structure or exposed to the weather. Some adhesives used for laminating interior members fail when exposed to the elements, and the resulting separation of laminations is both unsightly and hazardous to the structural framework of the building.

NECESSARY RESIDENTIAL DRAWINGS

The complete set of plans for a residence should generally include plot plan; floor plans; foundation (or basement) plan; mechanical (plumbing), air conditioning (heating), and electrical plans; exterior elevations; cabinet and interior elevations; details, sections, and whatever framing plans are necessary to explain the system.

The structural detailer in this instance would prepare those drawings required to explain the structural system, including framing plans, details, and sections. For a realistic example of what is required, look again at Figure 6-1 (western framing). The detailer would prepare *framing plans* for each floor and the roof. These plans would call out size, location, and direction of the various joists, beams, and rafters. The relationship of the various members would be further explained in *sections*, drawn to show again the various members, their direction, size, and spacing. Any items of important

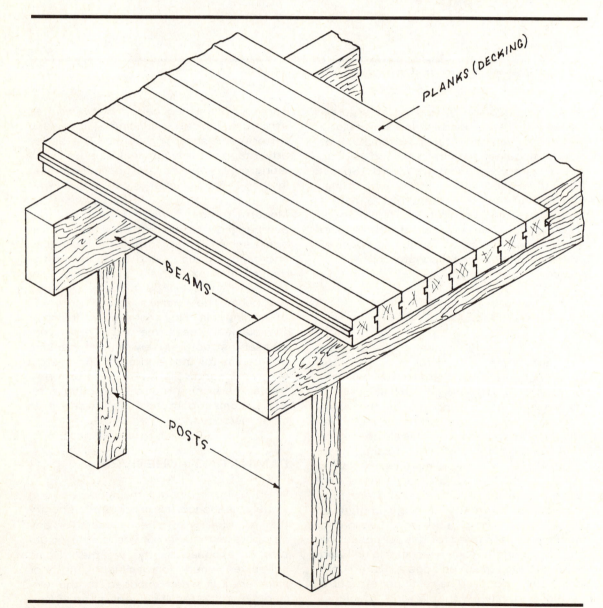

Figure 6·5 Plank-and-beam framing

information, or methods of construction joints, or members not shown (or not shown clearly) in the framing plans and sections would be delineated in *details* of the particular small area in question.

TYPICAL STRUCTURAL DRAWINGS—FRAME RESIDENCE

Combine the information of the isometric drawing of western framing (Figure 6-1) and the architectural floor plans in Figure 6-7, and observe the relationships of various structural components. Using these items for the rough configuration of the residence, the following drawings have been prepared to explain the structure of the residence.

Floor framing The detailer would initially observe that the first floor (lower or ground

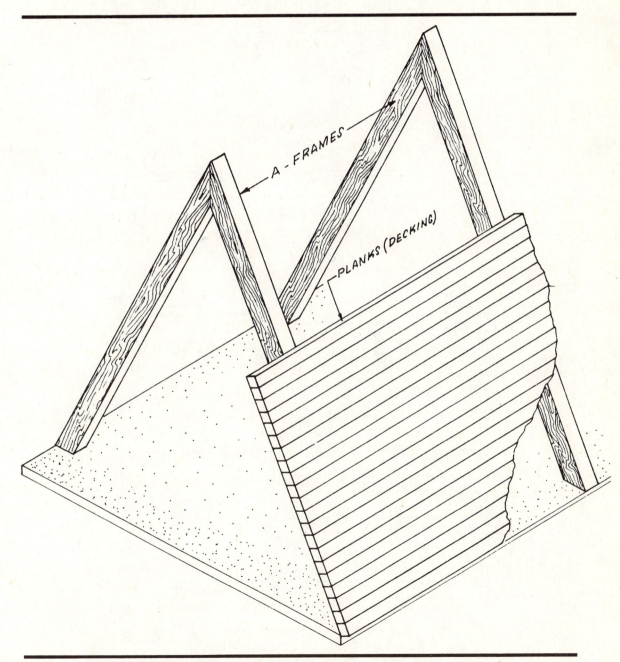

Figure 6·6 A-frame construction

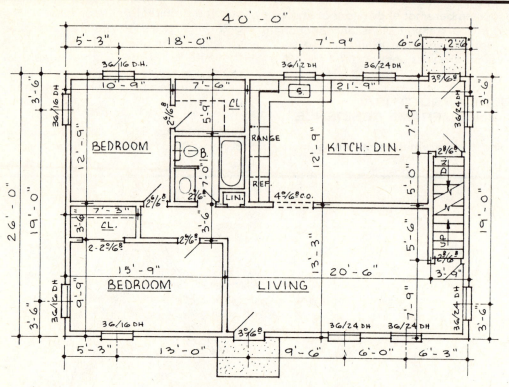

FIRST FLOOR PLAN

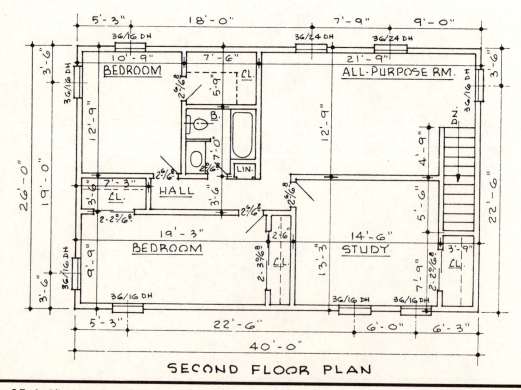

SECOND FLOOR PLAN

Figure 6·7 Architectural plans—first and second floors

floor) of the structure in Figure 6-8 was to be constructed using 2 × 10 floor joists at a spacing of 16 in. O. C. (on center, or center to center). In this particular case, the total width of the house is too great for a single span, so a girder must be run

through the center to provide bearing for the ends of the joists. Notice that the joists, girder, and other beams and framing are shown in their proper location by a single series of lines representing the individual pieces. These lines show the posi-

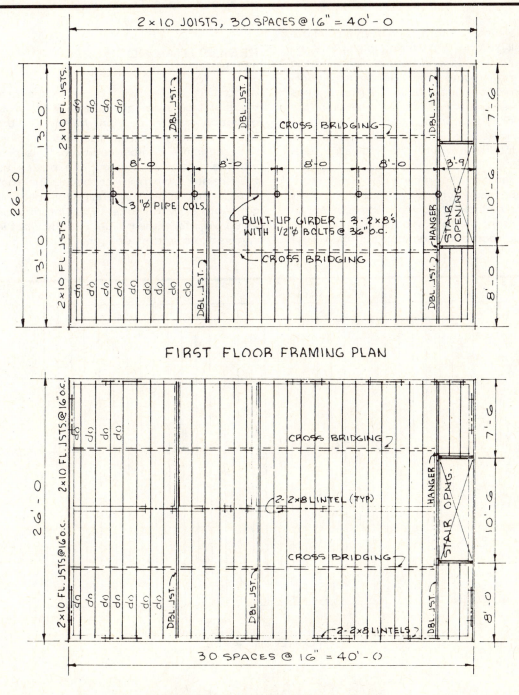

FIRST FLOOR FRAMING PLAN

SECOND FLOOR FRAMING PLAN

Figure 6·8 Framing plans—first and second floors

tion and direction of the members. The dimensioning is done in the standard manner, and the use of the laborsaving do (ditto) and the method of expressing cumulative distances as "number of spaces at (dimensioning) = total" are illustrated. Height from grade or finish elevation is noted, as with steel.

Supplementary information is given by note (or call-out) on the framing plan. The example for the second floor is similar to the first and indicates the total framing information for the upper floor. If a residence is observed just prior to the installation of the subfloor, the framing plan represents in single lines what the observer would see when looking down on the floor structure.

Roof framing The framing plan for the roof is again representing what an observer would see looking down on the roof structure before the roof sheathing is installed. For the roof, such items as subfacia (roof edge board), rafters, and ridge boards are shown as in Figure 6-9. It is helpful to note that when all roof pitches (the ratio of length of horizontal distance or *run* to amount of vertical distance or *rise*) are the same, the roof intersections (valleys and hips) occur at 45° angles with the roof edges and ridges. The example has a gable roof.

RESIDENTIAL ROOF TRUSS DETAILS

Wood trusses, as with those in the previous unit on steel, are generally drawn in elevation and in section. In section the truss is simply viewed from top, or side, with the initial member removed. This provides a means of viewing the pieces as they relate from front to back. The roof truss in Figure 6-10 is detailed for split-ring connectors with bolts. The information given will inform the builder exactly what is involved in the truss design.

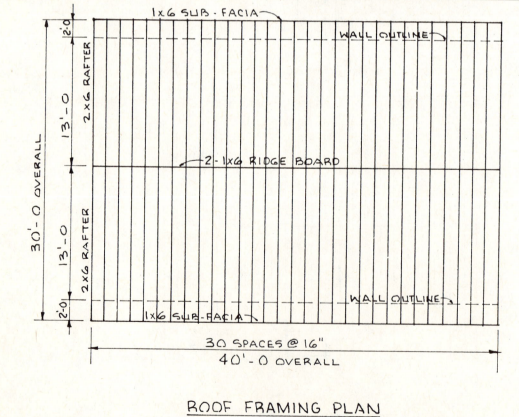

Figure 6·9 Roof framing plan

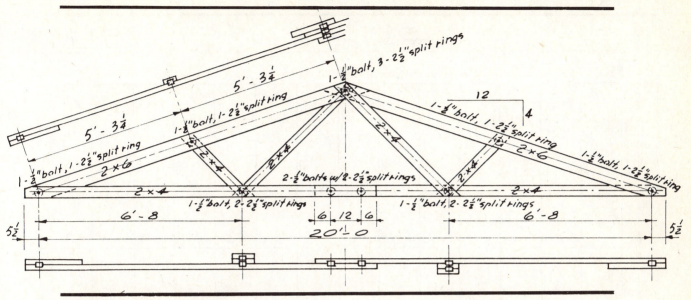

Figure 6·10   Detailed truss drawing

This truss is first laid out using the center lines of the members and centerline intersections at joints as the initial drawing step, with members then drawn showing their width to scale. To explain the truss fully, the example here includes front view (elevation), section below top chord, section above bottom chord, and the detail explaining the attachment of trusses at the top plate of the wall. The information noted on the drawing tells the method of fastening each joint, the size of each member, the dimensions of the truss, the roof pitch, the method and length of splices, and the amount of camber (if a factor).

Great care should be exercised in laying out the truss, and particular care should be given to explaining which members occur in front of or behind other members. Note that in front view the dotted (hidden) lines indicate one member passing behind another. In this case, the names of the various truss components are labeled. This is done frequently, but not always. Another variation in detailing the truss is that of assigning joints a letter symbol (A, B, etc.) and listing nailing or joint fastening conditions in an adjacent table (or schedule).

As with steel trusses, it is quite feasible to show only half of the truss if it is symmetrical about the center line. This practice has the disadvantage, however, of not providing the builder with the complete picture

of the truss and may result in incorrect lapping of truss members.

SAMPLE RESIDENTIAL DRAWINGS CATHEDRAL CEILING

The drawings for the residence shown in Figure 6-11 are typical of the necessary drawings for a structural system of this type. This particular residence has a basement and a single story. The basement floor and the upper walls are framed in a usual manner, but the ceiling is of the exposed-beam *cathedral* type, and the roof is supported by the exposed beams.

This application calls for composite roof beams built up of several pieces of dimension lumber. Notice the details explaining the beams and their connections. The ridge beams (main beams through the center of the residence) have end-bearing at the outside walls and at the stone fireplace. Since the spans are rather great, these beams are built up from two 2 × 12s with a ⅜-in. steel plate between them. Such a method of combining wood and steel components is far from unique.

POST-AND-BEAM DETAILING

Detailing post-and-beam construction differs from conventional framing plans in

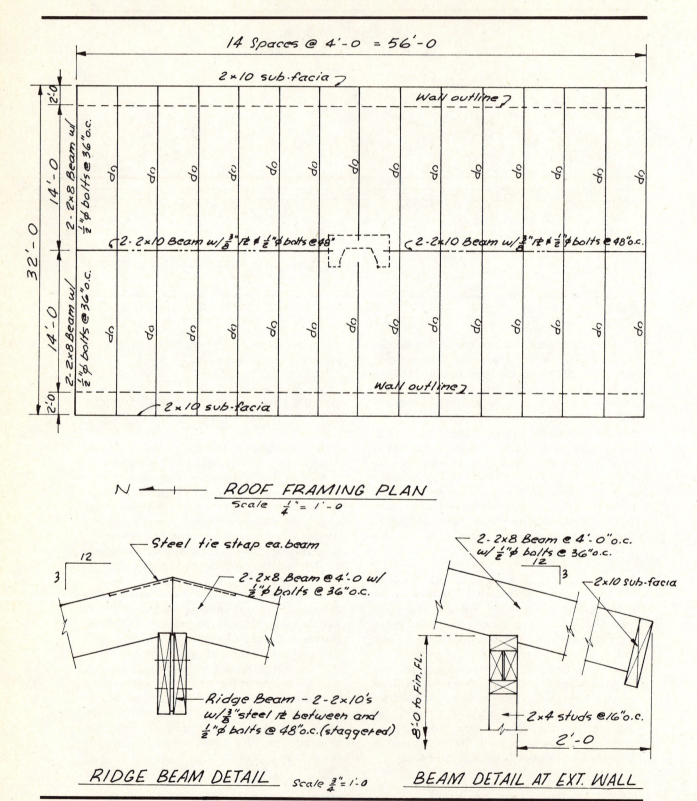

14 Spaces @ 4'-0 = 56'-0

2×10 sub-facia

Wall outline

2-2×8 Beam w/ ½"∅ bolts @ 36" o.c.

32'-0

14'-0

14'-0

2'-0

2'-0

2-2×8 Beam w/ ½"∅ bolts @ 36" o.c.

2-2×10 Beam w/ ⅜"℞ & ½"∅ bolts @ 48"

2-2×10 Beam w/ ⅜"℞ & ½"∅ bolts @ 48" o.c.

Wall outline

2×10 sub-facia

N

ROOF FRAMING PLAN
Scale ¼" = 1'-0

Steel tie strap ea. beam

12
3

2-2×8 Beam @ 4'-0 w/ ½"∅ bolts @ 36" o.c.

Ridge Beam - 2-2×10's w/ ⅜" steel ℞ between and ½"∅ bolts @ 48" o.c. (staggered)

RIDGE BEAM DETAIL   Scale ¾" = 1'-0

2-2×8 Beam @ 4'-0" o.c. w/ ½"∅ bolts @ 36" o.c.

12
3

2×10 sub-facia

8'-0 to Fin. Fl.

2×4 studs @ 16" o.c.

2'-0

BEAM DETAIL AT EXT. WALL

Figure 6·11   Exposed-beam roof details

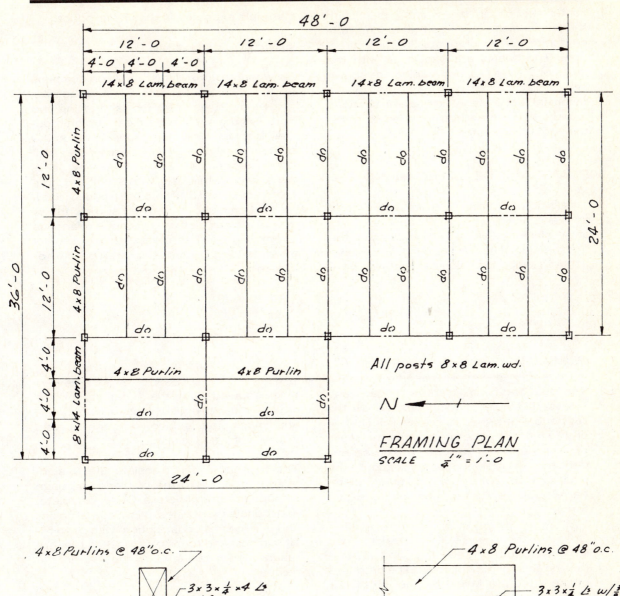

FRAMING PLAN
SCALE $\frac{1}{4}$" = 1'-0

All posts 8 x 8 Lam. wd.

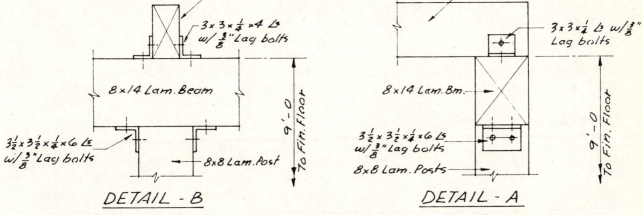

DETAIL - B

DETAIL - A

Figure 6·12 Beam-and-column framing plan and details

that columns (posts) are indicated in horizontal section on the framing plan, with their dimensions given as the spaces from center to center of posts (in each direction). The beams are shown as with the joists of Figure 6-8, and purlins are shown in a similar manner (see Figure 6-12). Attachment details and necessary sections round out the drawings required to detail this structural system.

PLANK-AND-BEAM DETAILING

This type of framing is represented as is post and beam, and the direction and size of the planks (decking) are indicated on the framing plan. The framing plan, in conjunction with joint and attachment details, delineates for the craftsman, the methods and materials expected in the final product.

SPECIAL SYSTEMS

No wood structural system is too complex, or too difficult to detail, if the detailer follows the rules of structural drafting and portrays the system in a clear-cut, methodical manner. He must understand the system, analyze its components, and simply represent its parts in plan view, sections, and details. As with steel and concrete structural drafting, the wood structural system must be detailed sufficiently to create a clear picture of it, but extraneous information and drawings should be deleted to avoid confusing the builder or erector.

SUMMARY

The post-and-beam, plank-and-beam, and stressed-skin-panel components allow the residential designer to create imaginative and functional structures within the realm of reasonable construction costs, as evidenced by residences throughout the country with roof designs of folded-plate, barrel-vault, butterfly, and other exotic forms.

One finds that the residential designer has immense latitude in selecting wood structural systems for his buildings. The products of the structural wood technology of our time, coupled with the creativity of the designer, can produce pleasant, functional, and imaginative solutions to the age-old problem of housing man.

The technician can translate the designer's intentions into the builder's language through structural details. Crisp, complete, and concise drawings allow for efficient and economical construction. Use of familiar symbols and terminology makes the representation meaningful to the man in the field.

Problems—Chapter 6

1. Draw a complete sidewall section and a complete endwall section, at the scale of ¾ in. = 1 ft 0 in., for the western framing shown in Figure 6-1, based upon the following:

   Height from first to second floor: 8 ft 1 in.
   Height from second floor to ceiling: 8 ft 1 in.
   Studs: 2 × 4 at 16 in. O.C.
   Floor joists: 2 × 10 at 16 in. O.C.
   Ceiling joists: 2 × 8 at 16 in. O.C.

Rafters: 2 × 6 at 16 in. O.C.
Roof pitch: 4/12
Subfloor: ½-in. plywood
Wall sheathing: ½-in. plywood

2. Draw a complete sidewall section and a complete endwall section (¾ in. × 1 ft 0 in.) for the balloon framing shown in Figure 6-2, based upon the specifications given above for problem 1.
3. The designer's sketches provide the information for a western-framed single-story residence with a crawl space. Use the information to prepare a floor-framing plan and a roof-framing plan at ¼ in. = 1 ft 0 in., and draw such details (¾ in. = 1 ft 0 in.) as you deem necessary for the structural system.

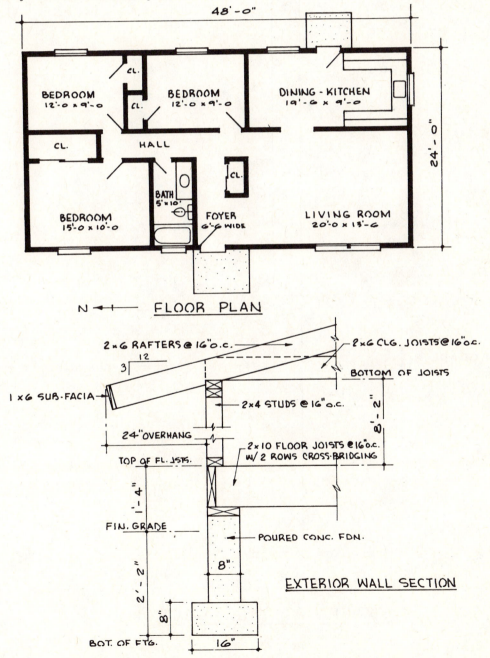

4. The sketch shown provides the necessary information to prepare details for the same single-story residence in problem 3, this time with a concrete floor slab. Draw the necessary structural wood drawings and details for the residence at a scale of 1 in. = 1 ft 0 in.

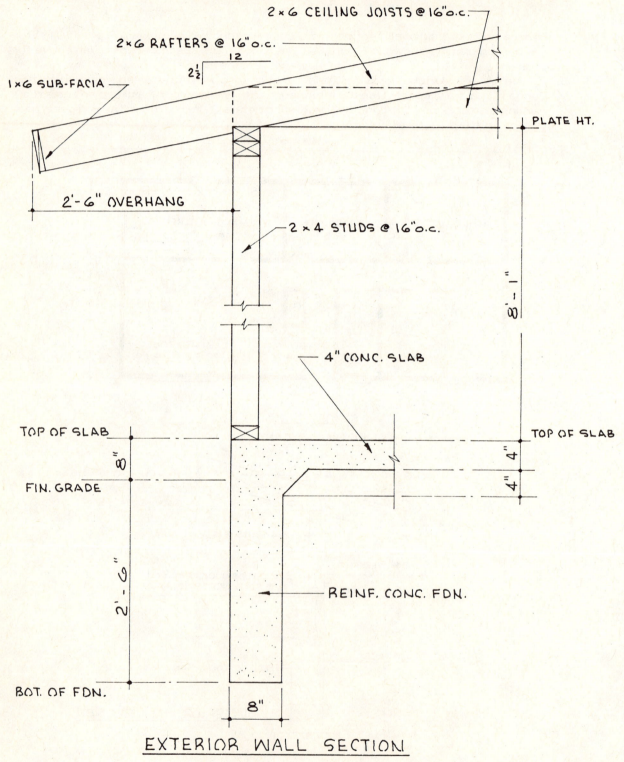

2 × 6 CEILING JOISTS @ 16"o.c.

2 × 6 RAFTERS @ 16"o.c.

$2\frac{1}{2}$ / 12

1 × 6 SUB-FACIA

PLATE HT.

2'-6" OVERHANG

2 × 4 STUDS @ 16"o.c.

8'-1"

4" CONC. SLAB

TOP OF SLAB

TOP OF SLAB

8"

FIN. GRADE

4"

4"

2'-6"

REINF. CONC. FDN.

BOT. OF FDN.

8"

EXTERIOR WALL SECTION

5. The plan in problem 3 and the detail sketch shown provide the information for a post-and-beam residence with 2½-in. wood roof decking. Prepare the framing plan and details for the residence. Be sure to detail the various connections required. The roof is to be a single-slope (shed) type, higher at the front than at the back. The top of the beam at the front is to be 10 ft 0 in.

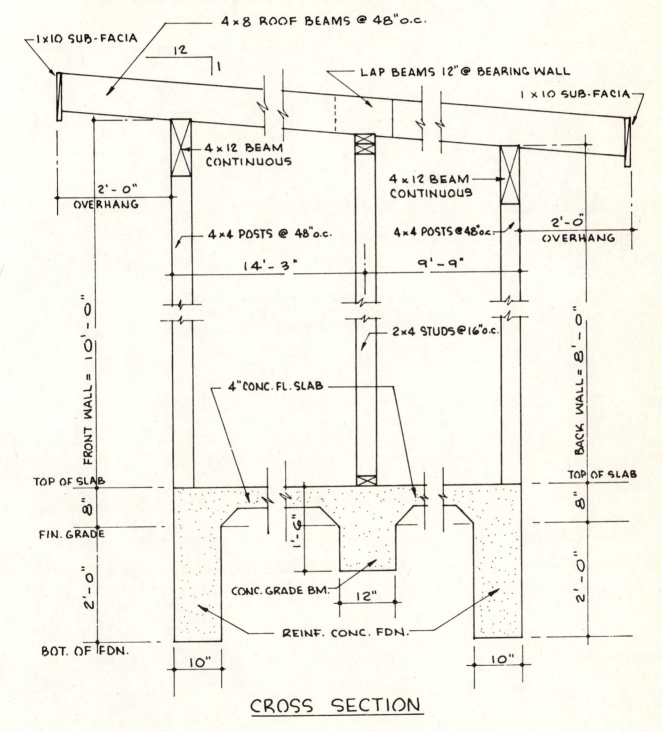

CROSS SECTION

6. Construct the necessary details to explain fully the roof truss shown in the rough center line sketch. The detail drawings should be prepared at a scale of ½ in. = 1 ft 0 in. Refer to Figure 6-10. (Show the complete truss, not just one-half of it.)

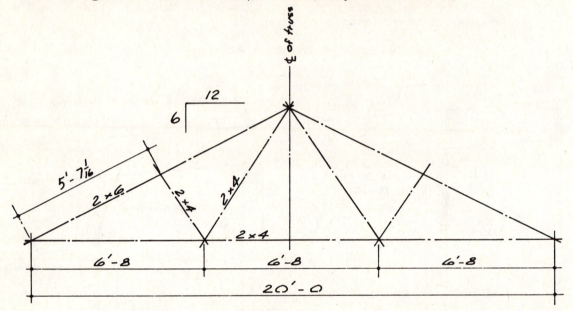

Note: Total Truss Length = 20'-11"
All Joints have ⅞"ø bolts and 2" split rings between joined members.

After Studying This Chapter You Should Be Able To:

1. Name the various framing systems used in commercial construction.
2. Define the term *camber*, and explain its function.
3. List the factors to be considered when detailing beam-end conditions.
4. Describe the effect of thermal expansion and contraction on beams.
5. Describe the various laminated wood beam shapes available.
6. Describe the direction and layout of individual laminations in the two most widely used lamination methods.
7. Name the two wood species most used for laminating wood structural members.
8. Describe the various types of wood decking commercially available.

INTRODUCTION

Designers have made great strides in recent years in making commercial structures less harsh and more inviting. Current efforts prove that the commercial building field has provided fertile ground for the work of creative architects. In the quest to provide more pleasant and humanized commercial structures, many designers have focused their attention on natural materials such as stone and wood. All types of commercial structures, including motels, restaurants, and churches, attest to the success of combining design creativity with the technology of wood construction.

Laminated wood products have overcome many of the limiting factors associated with wood structural components in the past. The use of laminated members easily

chapter 7
wood structural systems- commercial

9. Name the methods that can be employed to conceal electrical wiring in exposed wood structural systems.
10. Name the various arch and rigid-frame types currently available.
11. Name the type of foundation that is necessary for arches, and explain why it is often necessary to tie the arch legs together at the base.
12. Name several types of laminated wood trusses produced for the commercial structures market.
13. Define a folded-plate roof, and explain how stressed-skin-plywood panels can be used to advantage in such a system.
14. Tell how specific the detailer must be in detailing a new contractor-fabricated structural system.

permits the normal spans desirable in most commercial construction.

Although the prevalent residential framing types of western (platform) and balloon framing are not suitable for most commercial structures, the post-and-beam and plank-and-beam systems are often used. Wood trusses appear frequently in some types of commercial applications, and arches and rigid frames are widely used. This section addresses itself to the various wood structural systems used in commercial construction and to the practices and conventions used in detailing them.

LAMINATED BEAMS
AND CAMBER

Wall-bearing construction in commercial work is often used, and the combination of heavy masonry walls and laminated wood beams is observed in many instances. The beams in such a system may be *straight*, *tapered*, or of the *pitched*-beam type (see Figure 7-1). The straight or tapered beams may be used in flat or single-sloping (shed-type) roof designs. With proper connections, straight or tapered beams may be

joined in the center of the structure to form the basis for a gable roof.

The pitched (or symmetrically tapered) beam is one which has the roof slope built in, providing a high (or ridge) point in the center with a downward slope on each side. The bottom of pitched beams may be perfectly flat, but frequently it is sloped upward slightly toward the center to provide *camber*. Camber provides a means of permitting the beam to deflect (bend downward) under load conditions without the de-

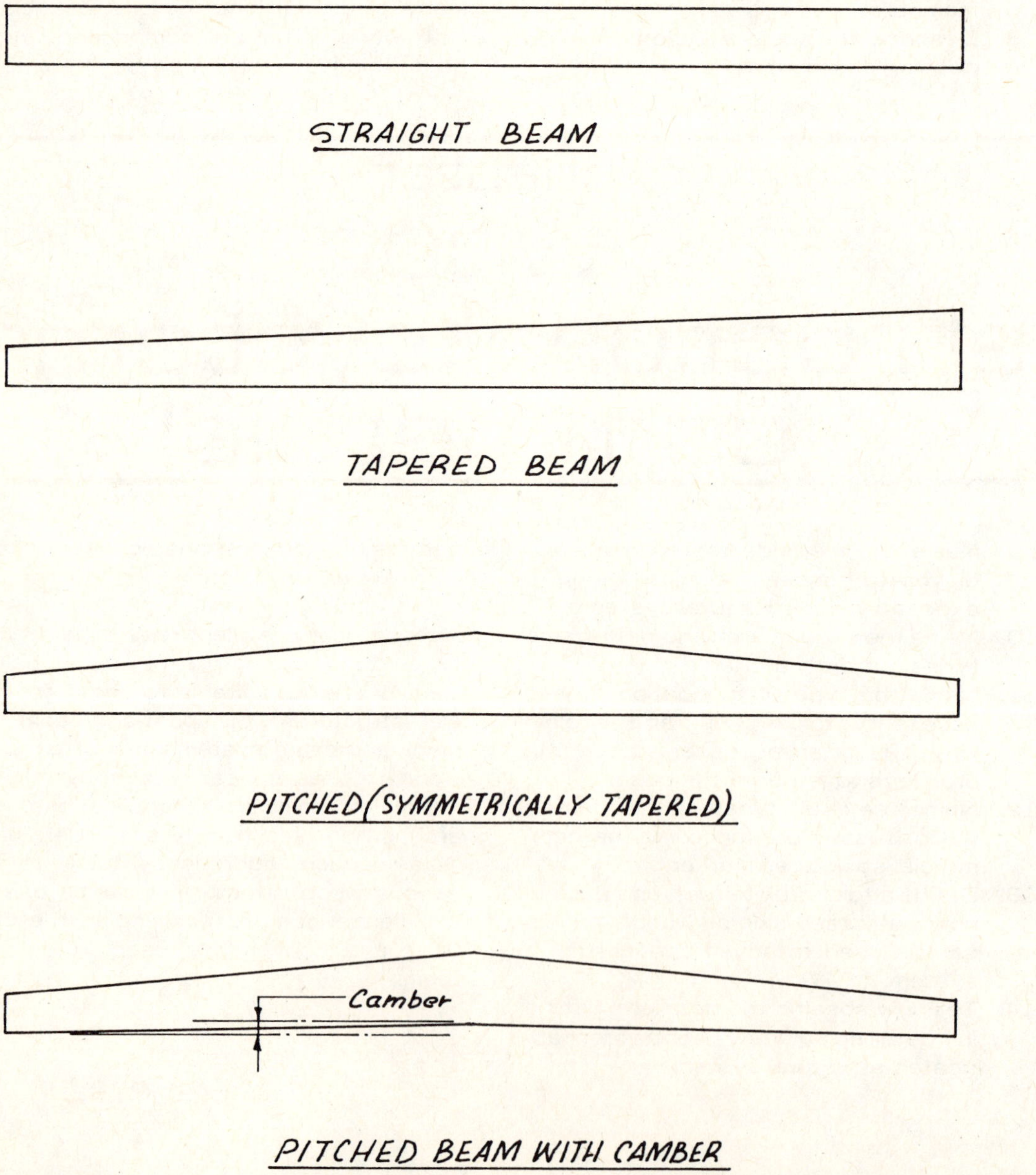

STRAIGHT BEAM

TAPERED BEAM

PITCHED (SYMMETRICALLY TAPERED)

Camber

PITCHED BEAM WITH CAMBER

Figure 7·1 Straight, tapered, and pitched laminated beams

flection being visible to the naked eye. Deflection of beams is generally limited to the range of $1/180$ to $1/360$ of the span. Most often the designer bases his calculations for the deflection to be limited to a maximum of $1/240$ of the span.

If a flat beam spans 60 ft, it could conceivably deflect 3 in. in the center quite safely. The practice of providing crown or camber would dictate that the beam be manufactured so that it was 3 in. higher in the center than at the ends. The deflection could then occur through loading, and the beam would assume the dead level position.

END CONDITIONS

The end (or bearing) conditions of beams (particularly in wall-bearing construction) are important and should be well designed and detailed. When extremely heavy loads are present, it becomes necessary to create pilasters (or thickened wall areas) below the points of contact. In some instances it is necessary to provide steel bearing plates to ensure that the supporting masonry work will not be crushed by the heavy concentrated load.

Extremely long spans create unusual conditions. Occasionally, because of the length of beams required for some large spans, expansion and contraction of beams due to temperature can create high stresses on the walls where the beams rest and are fastened. The stresses from thermal expansion tend to push the walls outward and apart. The contraction brought about through cooling of the mate-

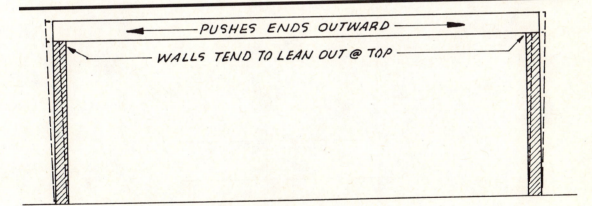

THERMAL EXPANSION

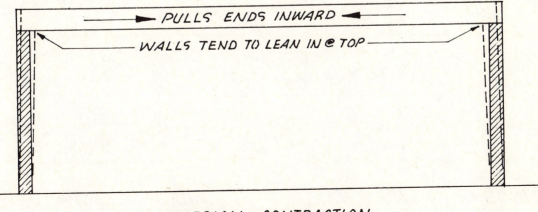

THERMAL CONTRACTION

Figure 7·2 Effects of beam movement on walls

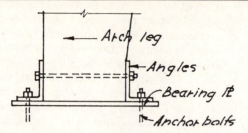

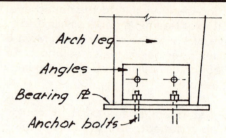

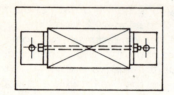

ARCH BASE DETAIL   ARCH BASE DETAIL

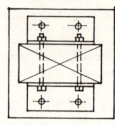

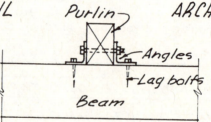

PURLIN DETAIL

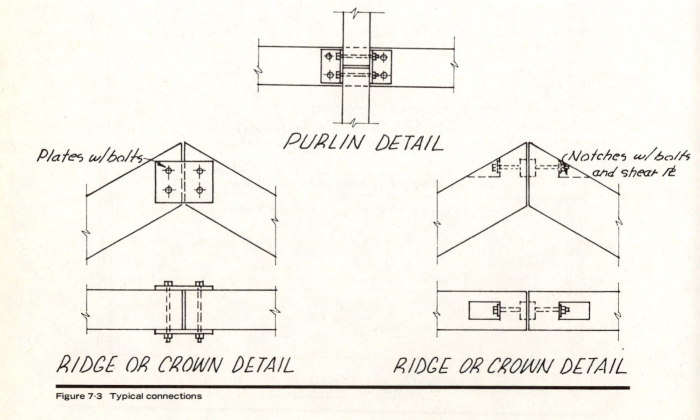

RIDGE OR CROWN DETAIL   RIDGE OR CROWN DETAIL

Figure 7·3 Typical connections

rial tends to pull the walls inward (see Figure 7-2). In extreme cases it becomes necessary to fasten only one end of the beam securely, while the other end is allowed to move freely as the beam expands and contracts. In some cases a Teflon-coated bearing plate is specified to reduce friction as much as possible.

CONNECTORS

Manufacturers of laminated wood components usually produce and recommend a number of different types of steel connectors. The connectors are available for splicing, joining beams and purlins, connecting arch sections at base and top, and for anchoring beams to walls and columns. Typical fastening details appear in Figure 7-3.

LAMINATION SYSTEMS

The laminations (individual strips) in a laminated beam or member are generally of the vertical or horizontal types (Figure 7-4). Notice that the grain of all laminations runs parallel to the length of the member. The most-used system of production relies upon the horizontal laminations.

BEAM MARKINGS

The actual practices followed in assigning marks to the various laminated components conform to the general guidelines used in working with steel. Beams might carry such marks as B-1, B-2, etc., or, even more specifically, marks labeled LB-1, LB-2, etc., referring to *laminated beam*. Some areas refer to laminated wood as *glu-lam*, and there may be other regional differences in terminology. Pertinent information regarding species, stress grade, and type of adhesive is often called out in tabular form or schedule on working drawings or shop drawings.

Many species of wood may be used to form laminated structural materials, but the two predominant species are douglas fir and southern yellow pine. These are the workhorses of the laminating industry.

EXAMPLE OF WALL-BEARING CONSTRUCTION

Figure 7-5 represents a floor plan for a small office building utilizing a wall-bearing construction system. The walls are of two layers of 4-in. brick, with laminated wood beams and purlins forming the roof structure. The roof deck of 2 × 8 ft insulating (composition) planks provides a prefinished ceiling surface as well. Notice that the beams are indicated with heavy single broken lines (similar to center lines). It should become clear from the plan that two

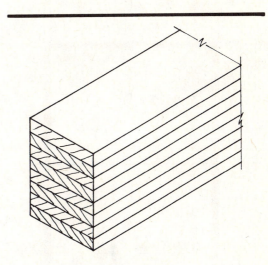

MOST-USED HORIZ. TYPE

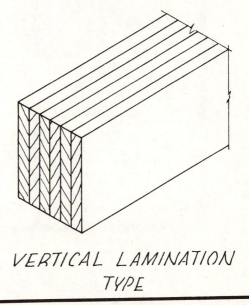

VERTICAL LAMINATION TYPE

Figure 7·4 Basic lamination types

different sizes of beams are necessary. Shop drawings should be prepared for each type of beam and a typical purlin. The framing plan in Figure 7-6 shows the components more clearly (along with their assigned marks), and the shop drawings and details in Figure 7-7 show the individual members. Notice that the shop drawings represent the beams in front view (full elevation) and call out holes to be drilled by the manufacturer, as well as notches, cuts, or other treatments necessary to be performed at the plant.

When purlins are combined with beams to form the roof structure (as in the building just discussed), there are two general methods used. If a uniform top surface level above beams and purlins is desired, the purlins must be framed between

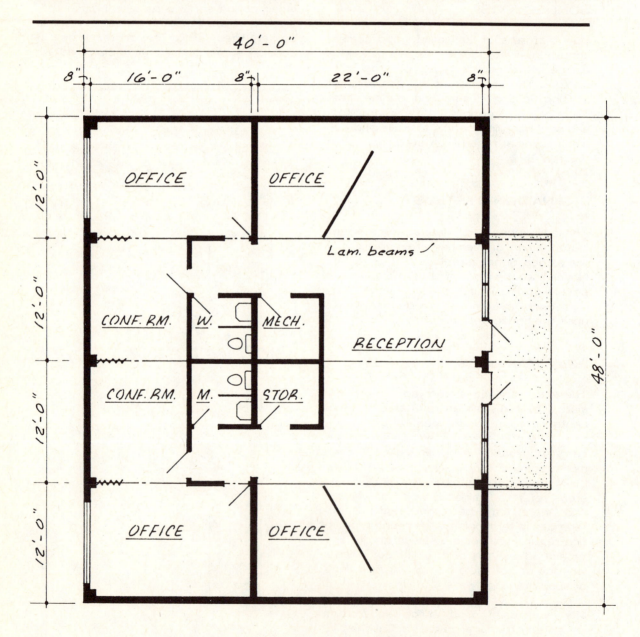

FLOOR PLAN

Figure 7·5 Wall-bearing floor plan

beams. This calls for some method of hanging or supporting the purlins between the beams (see Figure 7-8). The other method simply places the purlins on top of the beams, with regular fastening methods employed.

EXAMPLE OF PLANK-AND-BEAM CONSTRUCTION

A second example· of wall-bearing construction is that of plank-and-beam framing

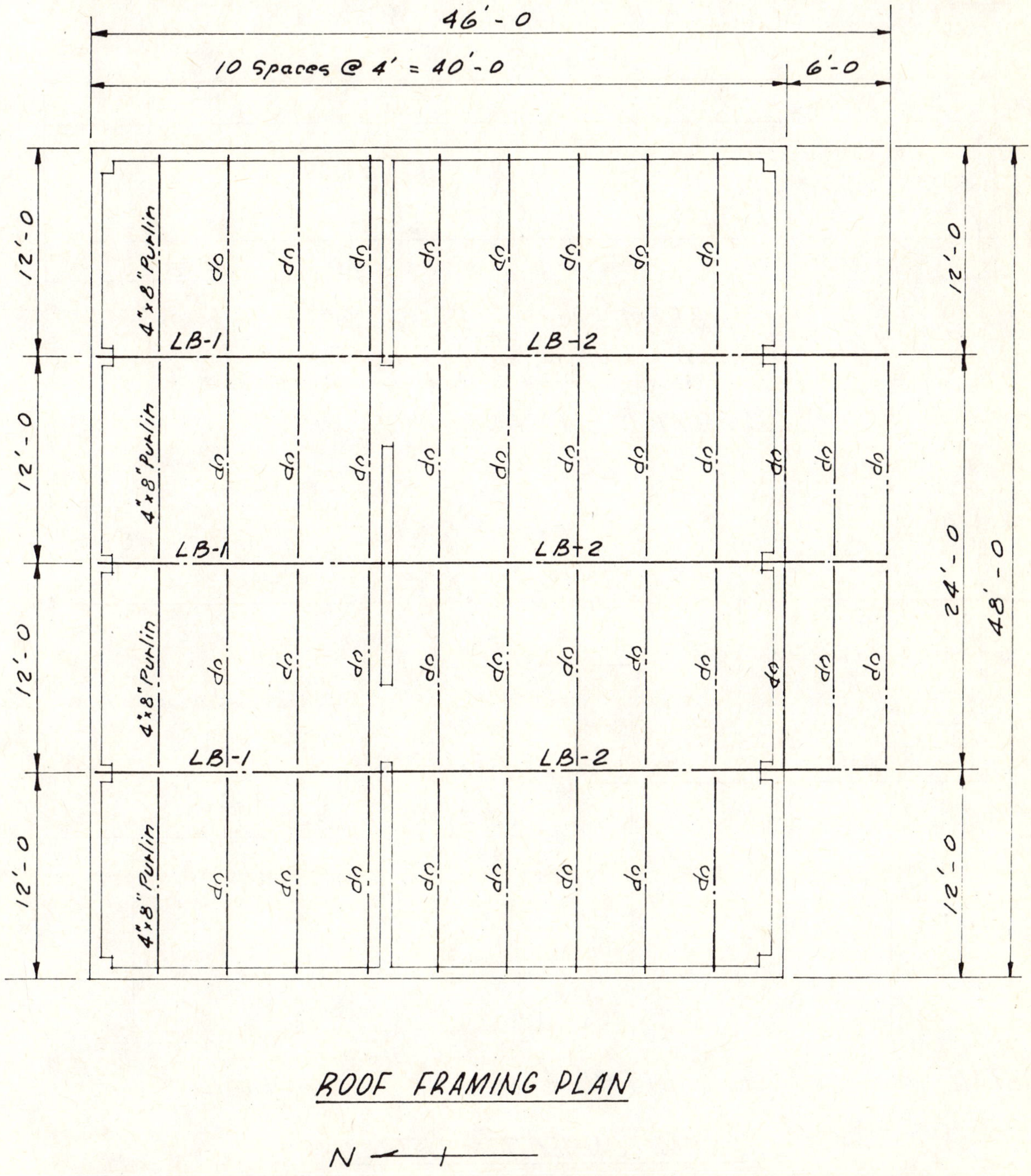

ROOF FRAMING PLAN

N

Figure 7·6 Wall-bearing framing plan

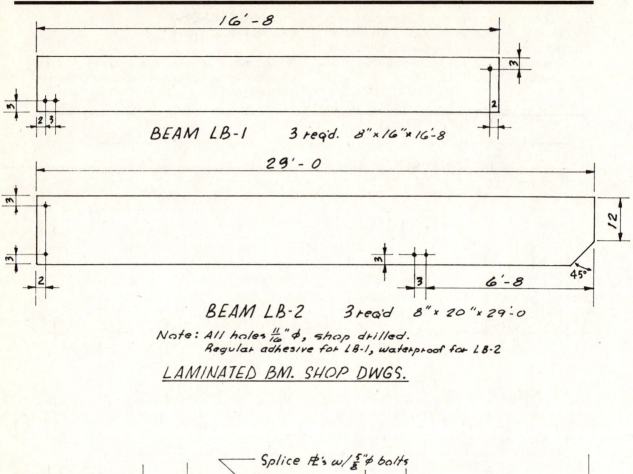

BEAM LB-1   3 req'd.   8" x 16" x 16'-8

BEAM LB-2   3 req'd   8" x 20" x 29'-0

Note: All holes $\frac{11}{16}$" ⌀, shop drilled.
      Regular adhesive for LB-1, waterproof for LB-2

LAMINATED BM. SHOP DWGS.

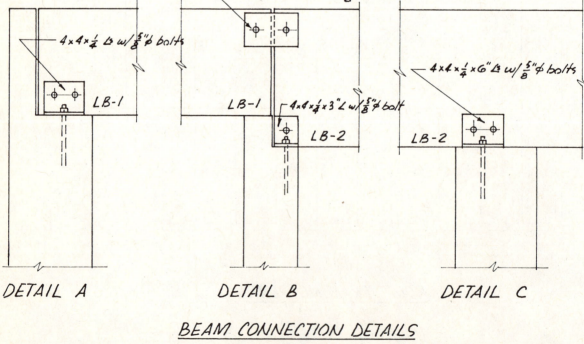

DETAIL A   DETAIL B   DETAIL C

BEAM CONNECTION DETAILS

Figure 7·7 Shop drawings and details

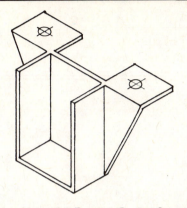

METAL HANGER FOR PURLINS

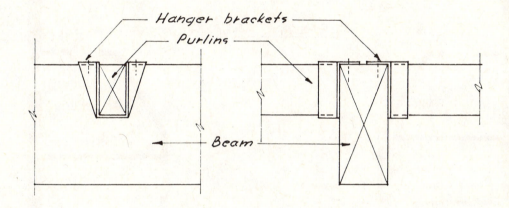

CONNECTION SYSTEM WITH HANGERS

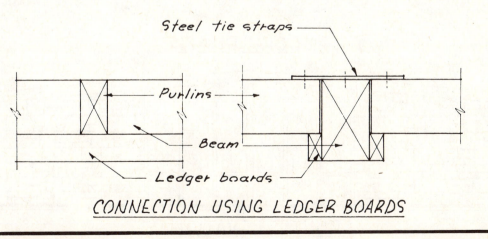

CONNECTION USING LEDGER BOARDS

Figure 7·8 Purlin connections

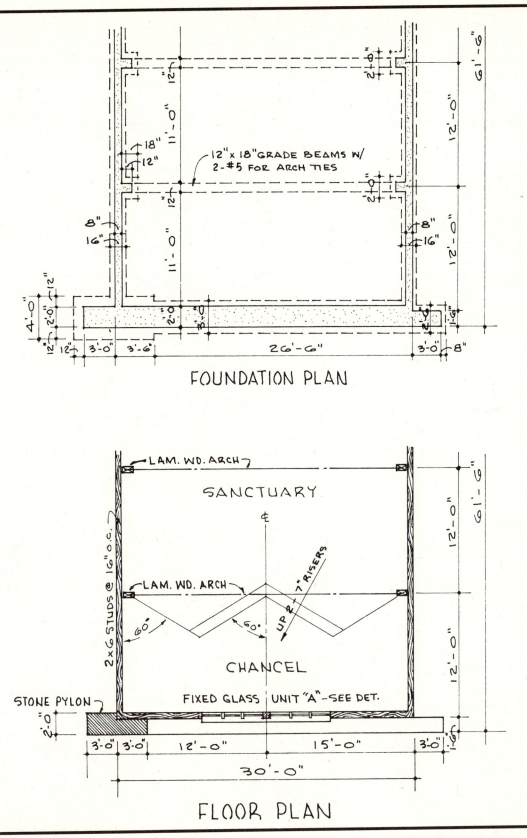

12" x 18" GRADE BEAMS W/
2-#5 FOR ARCH TIES

FOUNDATION PLAN

SANCTUARY

LAM. WD. ARCH

LAM. WD. ARCH

2x6 STUDS @ 16" O.C.

UP 2 - 7" RISERS

60°

60°

CHANCEL

STONE PYLON

FIXED GLASS UNIT "A"—SEE DET.

3'-0"   3'-0"   12'-0"   15'-0"   3'-0"

30'-0"

FLOOR PLAN

Figure 7·17 Partial church plans

bolts are actually welded to the reinforcing steel in the grade beam prior to pouring, thus forming a tension rod system.

A-FRAMES

Resorts and vacation cabins have made

the popularity of the A-frame most apparent. These pleasant structures can be erected quickly and easily, and the wide range of methods available for foundation support permit their use on practically any type of site. An A-frame is a form of the

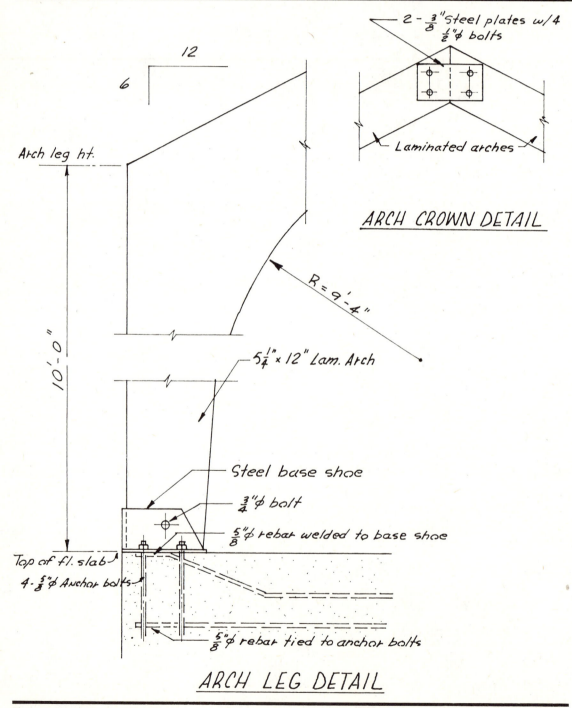

ARCH CROWN DETAIL

ARCH LEG DETAIL

Figure 7·18 Church arch details

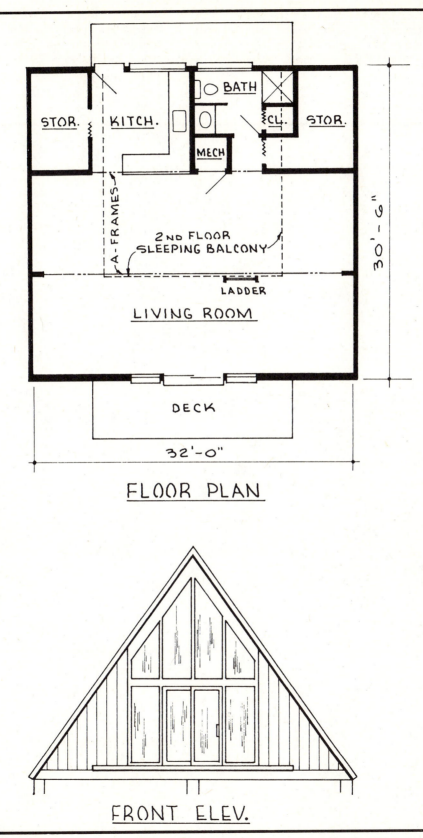

FLOOR PLAN

FRONT ELEV.

Figure 7·19 A-frame cabin

arch and permits an open interior devoid of columns or load-bearing walls. The A-frame vacation home in Figure 7-19 is typical of this type of structure. The details in Figure 7-20 delineate the method of construction and the individual members.

The student should immediately be aware that, in this example, the arches are tied together at the base by an additional laminated beam. The beam tying the arch heels together serves to support the floor decking and is placed on round con-

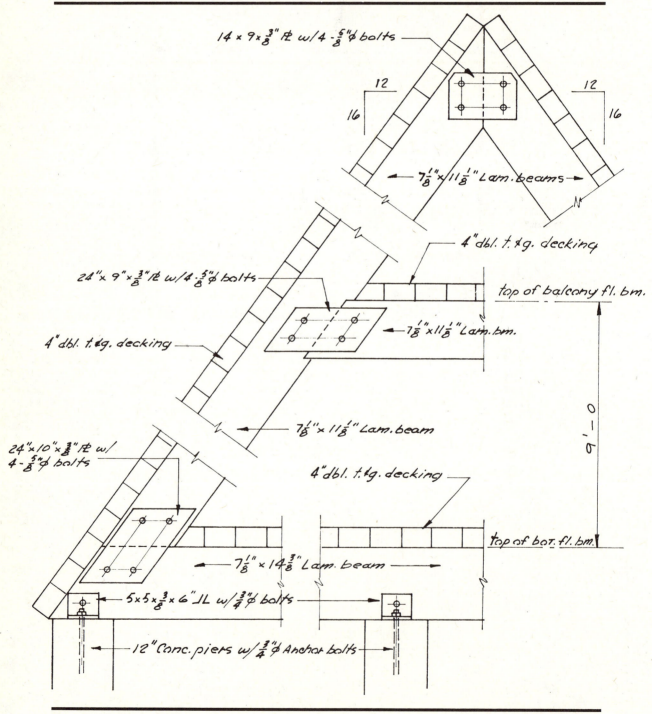

Figure 7·20   A-frame details

crete *piers* eliminating additional foundation work as well. With this particular configuration, it is apparent that such a structure could be erected over shallow-water shorelines and on steeply sloping mountain terrain.

The upper level of this cabin is provided by a second floor (with interior balcony rail) supported on secondary laminated beams bolted to the main A's forming the structural system. It is important to observe that the additional load of the second floor makes it necessary to use larger sections in the A-frames than would have been necessary for roof construction alone. Those members used in the area without the balcony floor could be smaller in cross section, but have been made the same for appearance factors and simplicity of construction.

The simplicity inherent in structural systems involving laminated wood arches and wood decking makes it feasible (and often mandatory) that architectural and structural details be combined. It would be inefficient and ludicrous to try to separate the items for two sets of details when each detail can show structural and architectural features and materials so clearly.

EXAMPLE USING STRAIGHT AND DOUBLE-TAPERED PITCHED BEAMS

Double-tapered pitched beams combined with 4-in. solid double tongue-and-groove wood decking form the roof structure for the sanctuary portion of the small church shown in Figures 7-21 and 7-22. The classroom and fellowship hall areas are based upon flat (straight) beams with the same decking. The wall construction is of hollow brick, which is a regular brick product except that the size of each unit is 8 × 12 in. and the centers have voids similar to concrete block for insulating purposes. Use of such brick permits a single-thickness wall which provides finished interior and exterior wall surfaces in one operation.

The heavy laminated beams are carried on integral pilasters laid with the hollow brick

walls, and the foundation and footings reflect the plan above. To tie the whole structure together (although the same side thrusts encountered with arches are not a significant factor here), a series of concrete grade beams was poured in the same pouring operation with the concrete floor slab.

The roof overhang was achieved easily by allowing the 4-in. decking to cantilever wherever possible and by allowing main beams to extend outside of the wall lines (or providing short supplemental beams) to support the decking where its direction paralleled the wall direction. The appearance and effect of the sanctuary interior is similar to that provided by arches, but the overall cost and the area subjected to wind load were significantly reduced through the use of the double-tapered pitched beams. Beam drawings as shown in Figure 7-23 were simple and were prepared to ensure that each laminated component was described and included.

A-FRAME FROM COMBINING STRAIGHT BEAMS

Straight beams, particularly those which are standard stocked items, are among the lowest-priced laminated components. These units can often be connected and combined to form interesting structural configurations without the increased costs of custom fabrication. A case of letting standard components serve in such a capacity is illustrated by the open-air chapel designed for a permanent scout camp. This small structure, shown in Figure 7-24, uses straight beams in A-frame form by simply job-cutting the ends in angular fashion and connecting them at the top. The use of the relatively light sections is further facilitated by the addition of light horizontal cross members for the A's themselves and light longitudinal members running between the A's. Laminated wood decking is used for the roof deck, topped with machine-split cedar-shake shingles. The small storage room has no tie structurally to the roof and A-frames, and has its own waterproof roof covering of translucent fiber glass.

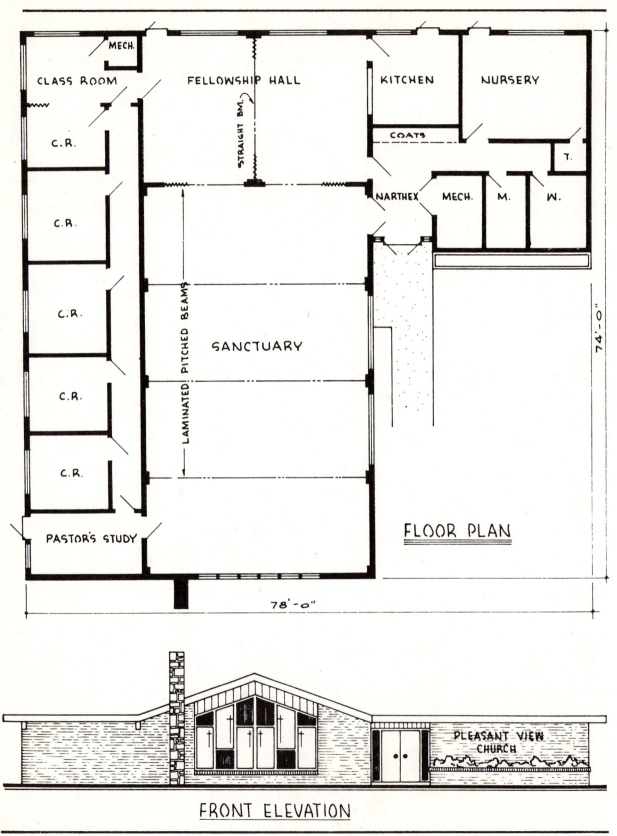

FLOOR PLAN

FRONT ELEVATION

Figure 7·21 Church preliminary drawing

The foundation of poured concrete serves the dual purposes of supporting the A-frames and supporting the concrete floor slab. No attempt was made to tie the legs of the A's together at the base, since the horizontal members tie them together at the 12-ft height.

Shop drawings are not necessary for standard billets (or beams), so in this case the details were combined with the architectural details (Figure 7-25), which clearly designated sizes, numbers, and the on-site operations necessary. Certainly the full all-weather exposure demanded waterproof adhesive for the laminated members, and further treatment was added on the job site to assure the proper weather-sealing of all materials.

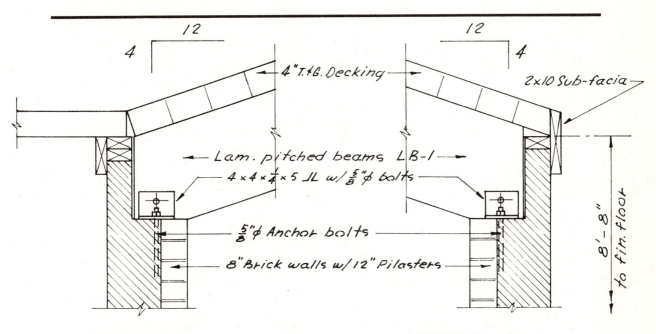

DETAIL A   DETAIL B

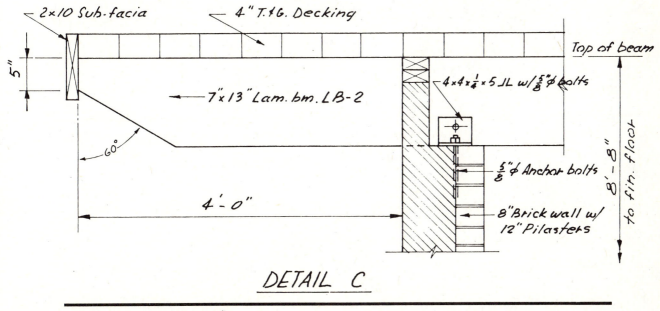

DETAIL C

Figure 7·22 Laminated beam details

LAMINATED TRUSSES

Although perhaps most used in industrial construction, the various laminated wood trusses are found in numerous commercial applications. These units are fabricated in the same manner as any other truss, except that the individual members are of laminated wood. The individual members are joined with bolts and heavy steel straps with shear plates added in those locations requiring them. Whereas many wood trusses based on regular dimension lumber are still fabricated with the individual pieces lapped, most laminated trusses have all members in the same plane.

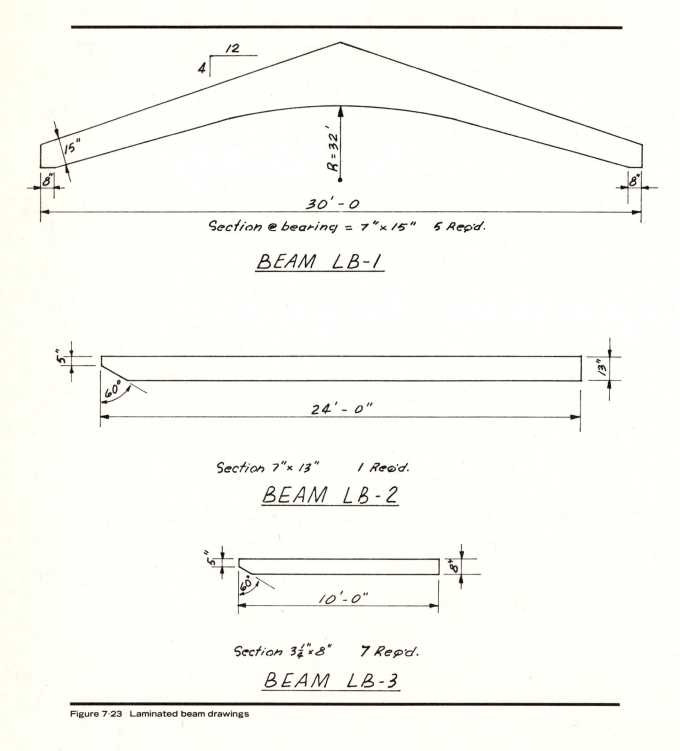

Figure 7·23 Laminated beam drawings

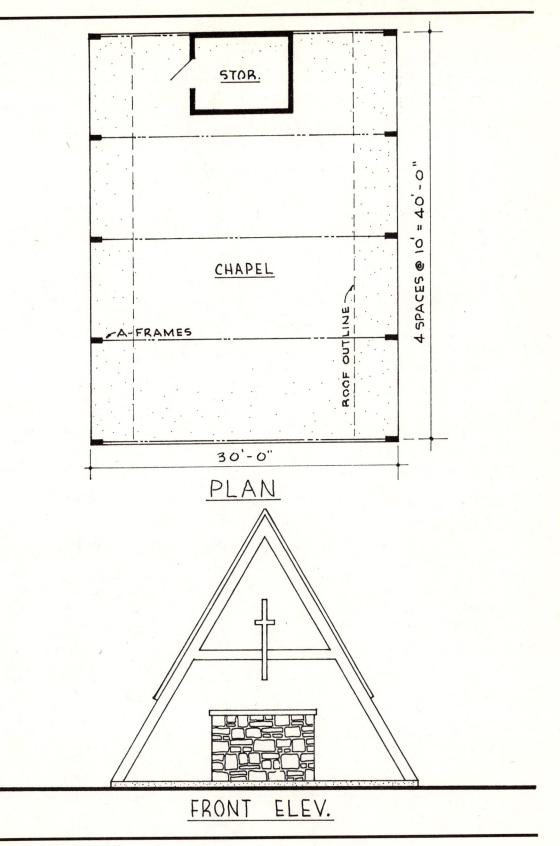

STOR.

CHAPEL

A-FRAMES

ROOF OUTLINE

4 SPACES @ 10' = 40'-0"

30'-0"

PLAN

FRONT ELEV.

Figure 7·24 Open-air chapel

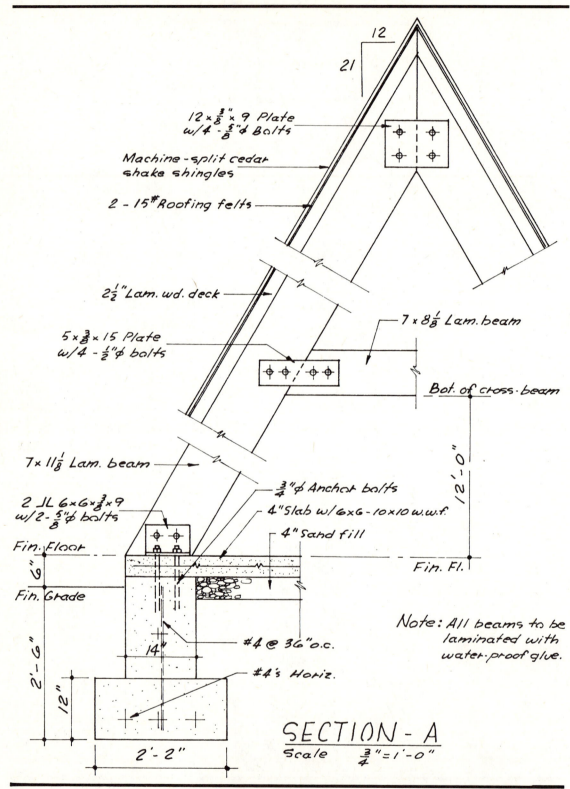

12

21

$12 \times \frac{3}{8}" \times 9$ Plate
w/4 - $\frac{5}{8}$"ϕ Bolts

Machine-split cedar
shake shingles

2 - 15# Roofing felts

$2\frac{1}{2}$"Lam. wd. deck

$7 \times 8\frac{1}{8}$ Lam. beam

$5 \times \frac{3}{8} \times 15$ Plate
w/4 - $\frac{1}{2}$"ϕ bolts

Bot. of cross·beam

$7 \times 11\frac{1}{8}$ Lam. beam

$\frac{3}{4}$"ϕ Anchor bolts

12'- 0"

2 ⅃L $6 \times 6 \times \frac{3}{8} \times 9$
w/2 - $\frac{5}{8}$"ϕ bolts

4"Slab w/6x6 -10x10 w.w.f.

4" Sand fill

Fin. Floor

Fin. Fl.

6"

Fin. Grade

Note: All beams to be
laminated with
water·proof glue.

2'- 6"

14"

#4 @ 36" o.c.

12"

#4's Horiz.

SECTION - A
Scale $\frac{3}{4}" = 1'-0"$

2'- 2"

Figure 7·25 Chapel details

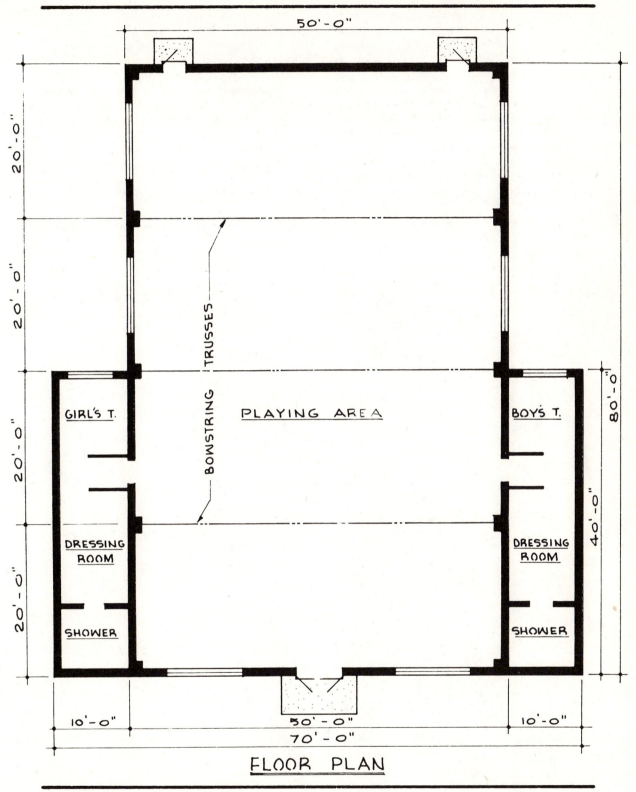

FLOOR PLAN

Figure 7·26 Elementary school gymnasium

Laminated wood bowstring trusses provide an efficient means for roofing large, open areas such as school and civic gymnasiums. The elementary school gymnasium shown in partial plan in Figure 7-26 is designed using concrete block load-bearing walls, with exterior brick veneer and bowstring laminated wood trusses. As with regular laminated beams, the walls must be made thicker with pilasters to accept the highly concentrated loads transmitted at the bearing points of the trusses. Details appear in Figure 7-27.

Factory-fabricated laminated trusses are most often preengineered and may in fact be a standard production item. This precludes the necessity for shop drawings as such, but the detailer frequently prepares a less-detailed elevation of the truss to assure the proper overall dimensions. Any public structure where the trusses are exposed to view (as in the example) dictates that camber be built into the bottom chord to avoid visible deflection. Several of the various types of laminated trusses available are illustrated in Chapter 8.

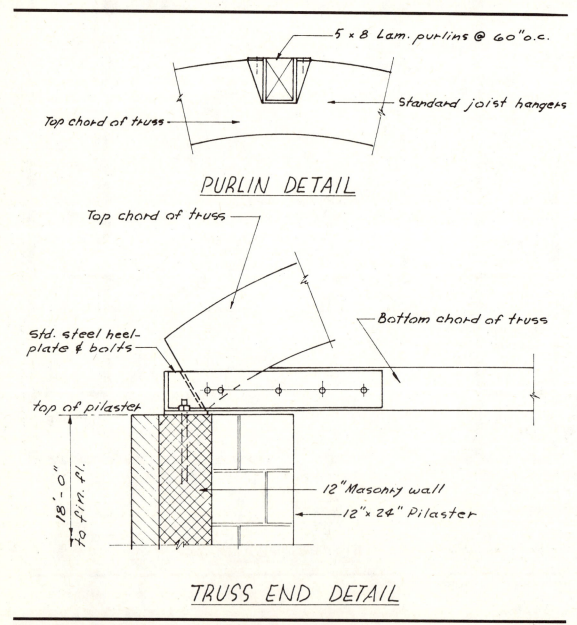

Figure 7·27 Gymnasium roof details

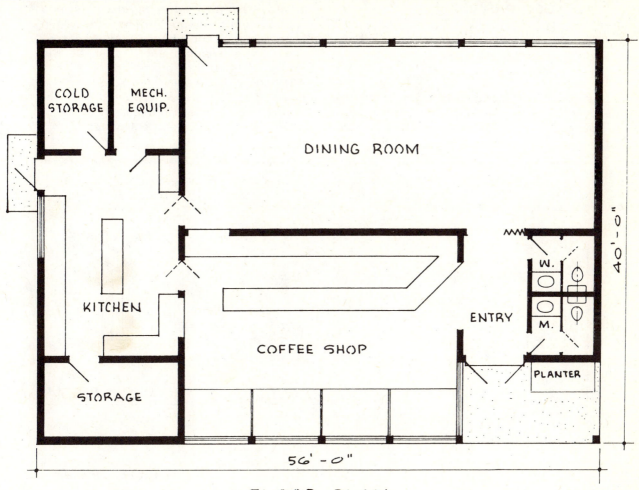

FLOOR PLAN

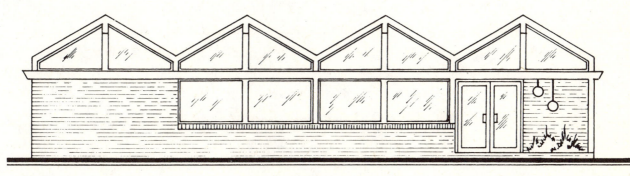

FRONT ELEVATION

Figure 7·28   Restaurant with folded-plate roof

FOLDED PLATE WITH PLYWOOD PANELS

Chapter 6 mentioned a variety of prefabricated plywood panels and components. These components frequently find use in small commercial structures. The small restaurant building shown in Figure 7-28 was constructed using large prebuilt plywood panels for forming the folded-plate roof

system. The panels combined the structural system, finished ceiling, insulation, and roof deck into one unit. From the details in Figure 7-29 it should become apparent that the panels support each other at the ridges and that they are reinforced by the addition of the beam at the valley. The ridge beam is more decorative than functional, and provides an excellent opportunity to conceal the electrical system.

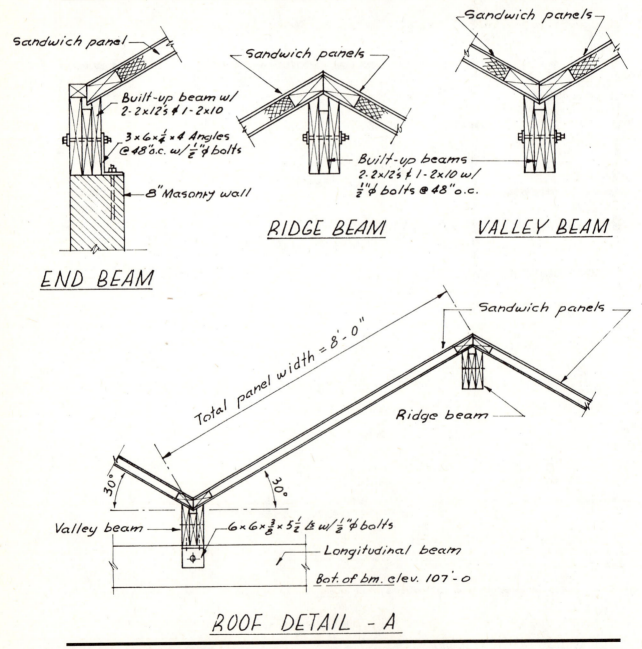

Figure 7·29 Folded-plate details

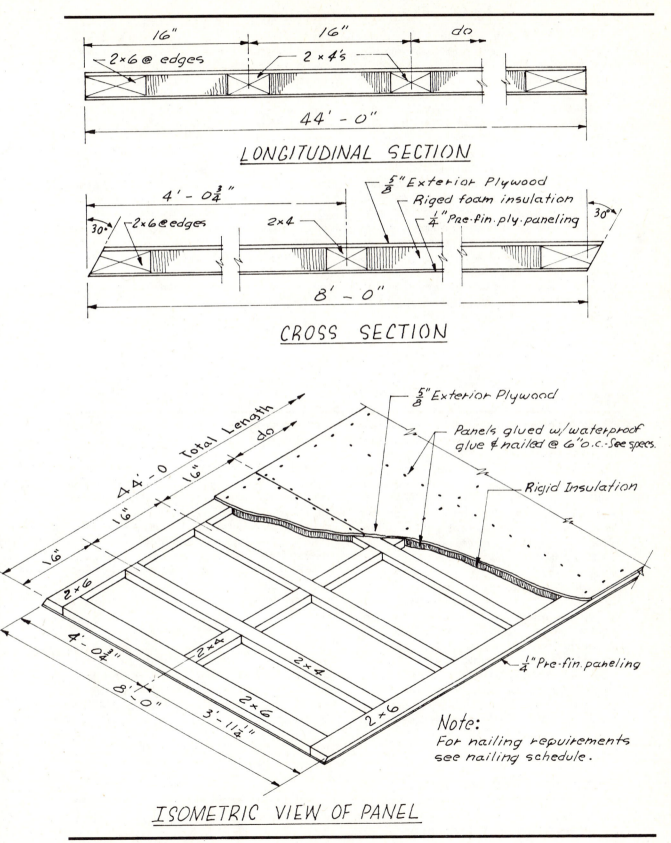

Figure 7·30 Roof panel sections

The fact that the roof panels were contractor-built for the job made manufacturer's shop drawings unnecessary; however, extensive drawings were necessary to guide the contractor in fabricating the panels prior to trucking them to the site. The beams are shown in the regular fashion on the floor plan. For this structure the roof panels were detailed once as regular sections and a second time as panel construction details only. Figure 7-30 shows the contractor how to fabricate the panels. Notice that, in addition to regular detailing practices, pictorial (isometrics in this case) drawings were prepared to make it easier to visualize exactly what the designer's intentions were. Many structural detailers confine their efforts to the orthographic area exclusively, but use of simple pictorials occasionally is more than justified to explain complex joints, cuts, connectors, and unconventional construction details.

The beams in this particular project were built up by combining several pieces of regular dimension lumber. These units were again extensively detailed to explain fully what was intended. The load-bearing walls of the structure were constructed of structural glazed tile (interior) and an exterior veneer of glazed face brick. A simple cap (or wood plate) was not used for this structure, but all walls were topped with a built-up wood beam. Since this was a departure from the more conventional practices, details (as shown in the illustration) were prepared designating the procedures and materials necessary for this application. All beams on this structure were fabricated from several separate pieces bolted together. For the design effect, square telephone-pole washers were used in lieu of standard washers.

SUMMARY

Countless commercial building designs and methods exist which have relied upon wood as the prime structural material. Wood, whether in solid or laminated form, imparts a natural beauty and warmth to building exteriors and interiors. The wide variety of wood structural products permits the architect a great amount of latitude in arriving at creative, functional solutions to building design problems.

As with any structural material, wood systems must be sufficiently well detailed to provide adequate information for contractor, supplier, and fabricator. Basically the same practices followed in detailing structural steel are applicable to detailing wood structural systems. Prefabricated and laminated wood structural components are sometimes detailed by the designer and his staff, while the actual shop drawings are often left to the product manufacturer.

Commercial applications for wood structural systems can utilize (but are not necessarily limited to) laminated beams, arches, trusses, plywood sandwich panels, and units created by combining various pieces of regular dimension lumber. Wood lends itself to wall-bearing, post-and-beam, and long-span structural methods. Many types of standard steel connectors are available for joining wood structural members.

The entire area of commercial building has provided the creative architect with numerous opportunities to use wood innovatively as a structural material. The results of such innovation clearly point to the flexibility and adaptability of the material.

Problems—Chapter 7

1. Prepare the details and shop drawings for the three beams shown.
   (see the diagram on the next page)

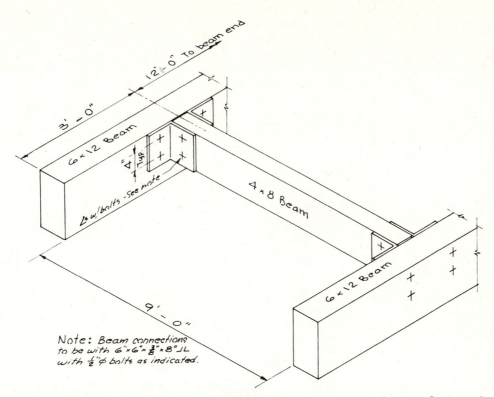

Note: Beam connections
to be with 6"x6"x⅜"x8"JL
with ½"∅ bolts as indicated.

2. Draw the design plan, connection details, and shop drawings for the four-post, four-beam assembly shown in the drawing.

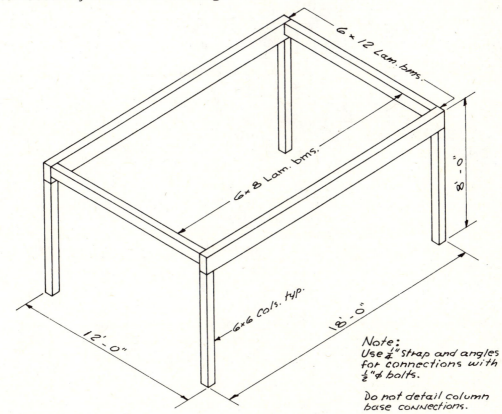

Note:
Use ¼" strap and angles
for connections with
½"∅ bolts.

Do not detail column
base connections.

3. The typical section illustrated is for a small office building with wall-bearing construction. Prepare connection details and shop drawings for the laminated beams and purlins. Note that the purlins are suspended between the beams.

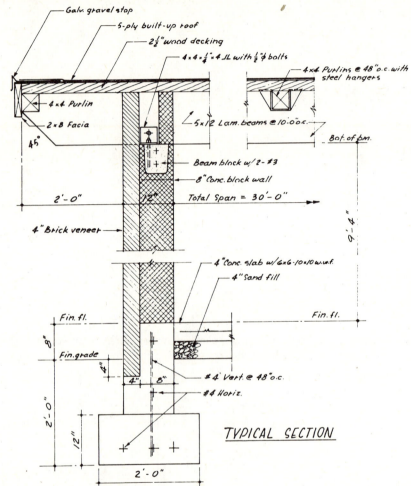

TYPICAL SECTION

4. Prepare details and shop drawings for the s` utilizes a tapered laminated beam and lamin·

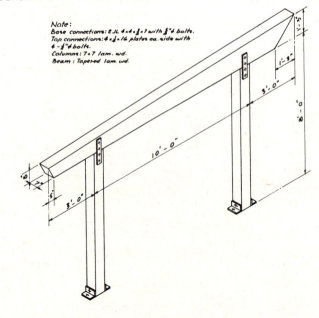

Note:
Base connections: 2 JL 4×4×⅜×7 with ⅝"ø bolts.
Top connections: 4×¼×16 plates ea. side with 4 - ⅝"ø bolts.
Columns: 7×7 lam. wd.
Beam: Tapered lam. wd.

5. The small open-air chapel shown relies upon laminated wood A-frames and decking for its structural system. Use the information given, and prepare a framing plan and details (refer to the example in this chapter).

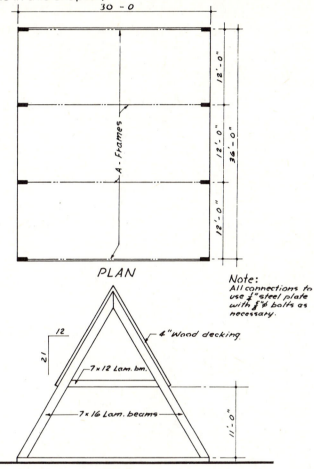

Note:
All connections to use ⅜" steel plate with ⅝"ø bolts as necessary.

PLAN

ELEV.

6. Draw the framing plan and connection details for the structure shown in the illustration.

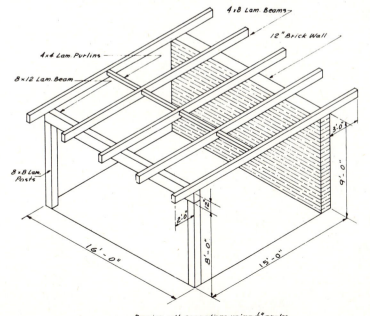

Design all connections using ¼" angles and plates as necessary, with ⅝"ø bolts.

7. The freestanding cross is built with laminated wood sections. Prepare the necessary details and shop drawings for its construction and erection.

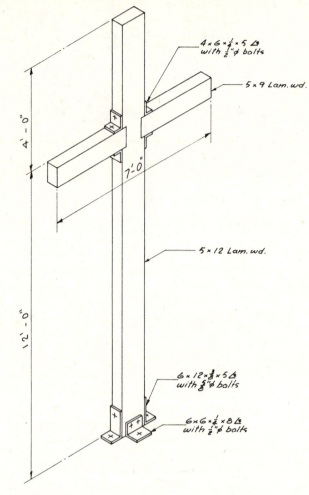

**After Studying This Chapter
You Should Be Able To:**

1. List some types of installations in which wood structural systems would be more advantageous than other types.
2. Describe some of the laminated wood products which lend themselves well to industrial construction.
3. Name the conditions that industrial structural systems are subjected to, which are not usually encountered in commercial buildings. Explain how they can be counteracted.
4. List some of the factors which influenced the development of mill construction.
5. Describe two main types of structural forms which typify mill construction.
6. Describe the various metal components used in mill construction.

14. Explain why it is necessary to provide bracing between trusses in large industrial structures.

INTRODUCTION

Early industrial structures relied upon wood as a structural material. Such structures are still being constructed using wood structural systems. Light steel structures with prefabricated components are in widespread use for small industrial buildings, but certain industrial functions preclude the use of steel for structural members.

Some industrial processes involve chemical elements which are extremely corrosive and create conditions where rusting of metal components is constantly prevalent. Wood members may provide the solution to

chapter 8
wood structural systems-industrial

7. Explain how the ends of beams are attached to bearing walls in mill construction.
8. Cite the advantages which laminated members offer when used in standard mill construction.
9. Say whether wall-bearing, post-and-beam, and long-span structural systems are used in industrial buildings. If so, cite examples.
10. Describe a tied arch, and explain how it differs from a regular radial arch.
11. Describe a hinge connection for arches, and explain why it is necessary.
12. List some of the metal fastening devices used in attaching beams to posts, posts to foundations, and arches and trusses to bearing walls.
13. Describe the following items and give their purposes: (*a*) lateral ties, (*b*) ledgers, (*c*) shoes, (*d*) corbel, and (*e*) diagonal and longitudinal bracing.

many of the problems of permanence and maintenance in such situations.

Wood, as mentioned in previous chapters, is easily worked with at the job site and can be erected in satisfactorily short periods of time. Heavy wood structural systems can carry acceptable fire ratings because of the ability of members to char on their exterior surfaces for fairly long periods while maintaining their load-carrying capabilities. Fire is a prevalent danger in any industrial structure, and exposed steel (unlike wood) fails suddenly and completely upon reaching its critical temperature.

Industrial structures often demand large, open space devoid of interior columns and walls. Certainly such buildings as aircraft hangars, warehouses, manufacturing plants, and processing facilities can make excellent use of long-span framing systems. The development of laminated wood components and systems has simplified the methods necessary to provide large, open space, and has been instrumental in increasing the use of wood as a structural material for industry. Laminated beams, arches, trusses, and domes provide an economical means of spanning large distances.

The functional aspect of any industrial building is of prime importance. The building must fit the purposes for which it is intended. It must also stand up well under the usage it receives. Heavy mobile equipment is often found in the buildings of industry. Items of equipment such as trucks, tractors, forklifts, etc., constitute a constant threat to structural members. Heavy material in storage or in production can exert sizable pressure upon the structure. Adequate design of structural members, joints, and connections is imperative in the industrial setting.

Function (as mentioned previously) must guide the designer's hand, but modern wood technology has provided the means of creating more attractive industrial structures without sacrificing the functional attributes. There is a natural attractiveness about laminated structural members, and solid or laminated wood decking creates a pleasant ceiling when exposed to view.

MILL CONSTRUCTION

The type of wood or timber construction known as mill construction evolved from early efforts to construct industrial mills, railroad train sheds, and factories. Through frequent use, a relatively standard system was established, as far as the various connections, joints, and details were concerned. Many of the details remain unchanged today.

The overall configuration of mill types was predicated upon the need for large spaces, adequate natural light, and adequate ventilation to protect workers from the fumes and smoke generated by the work within such buildings. Essentially two main types of structures represent most mill construction.

Main mill construction types. The *gabled-roof* type may have a raised section in the center for light and ventilation, and may also have lower bays at one or both sides of the main section. The *lumber-mill* type of structure creates the clearly identifiable sawtooth roof pattern used on many factory structures, again providing natural light and the opportunity for adequate ventilation.

The sketches in Figure 8-1 illustrate some of the variations of the gabled-roof and lumber-mill types of mill construction. Such construction may be multistory.

Mill construction components and methods The mill construction system often combines heavy load-bearing masonry walls with heavy lumber floor and roof construction. In general, the ends of the heavy timber beams bearing on the walls are set into recesses in the walls. The beam ends rest on cast-iron wall plates (bearing plates) or in cast-iron or steel wall (or hanger) boxes. The plates or boxes are fabricated with integral anchor lugs to facilitate anchoring them to the masonry work.

Posts (columns) are usually set on cast-iron bases which have been anchored to the concrete foundation. The base may consist of a flat steel plate with angles or straps for fastening the post, or the base may be a cast (or fabricated) unit which totally encloses the bottom of the post. The tops of posts (columns) are attached to beams with steel or cast-iron caps. The caps (as the bases) may enclose the top of the post, or be a unit made up of plate and angles or straps.

Mill construction often calls for posts (columns) to run for several stories. In such cases, provisions are made for metal bearing surfaces between the top and bottom of the two posts. Steel plates (with or without brackets) are used for this purpose. Cast-iron *pintles* and bases are often used to transmit stresses through such columns. The pintle is a cast-iron unit which serves as the base of the upper column. The pintle rests, in turn, on the cast-iron cap of the column below. The base and

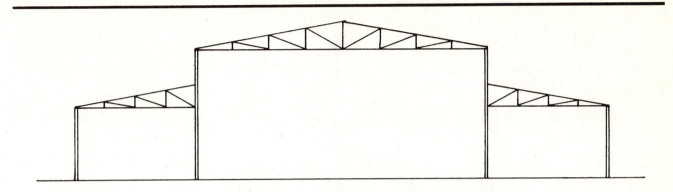

GABLE ROOF TYPE WITH SIDE STRUCT.

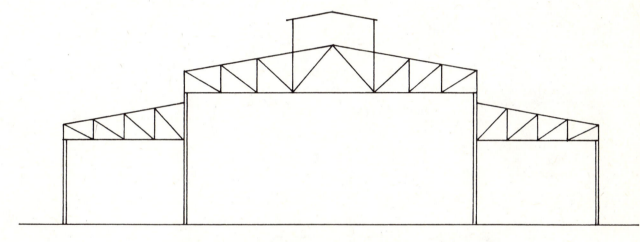

GABLE ROOF TYPE WITH CENTER MONITOR

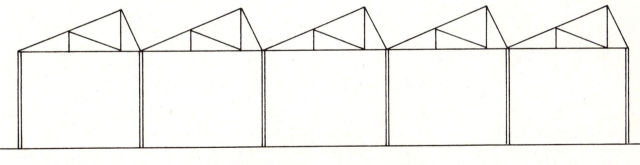

SAWTOOTH ROOF TYPE

Figure 8·1 Typical building shapes for mill construction

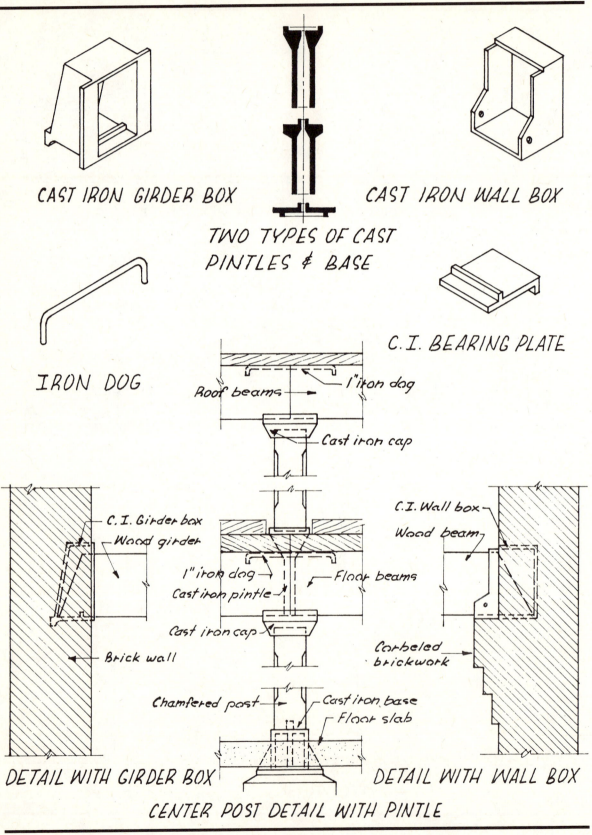

CAST IRON GIRDER BOX

CAST IRON WALL BOX

TWO TYPES OF CAST
PINTLES & BASE

IRON DOG

C.I. BEARING PLATE

Roof beams

1" iron dog

Cast iron cap

C.I. Girder box

Wood girder

1" iron dog

Cast iron pintle

Cast iron cap

Brick wall

C.I. Wall box

Wood beam

Floor beams

Corbeled brickwork

Chamfered post

Cast iron base

Floor slab

DETAIL WITH GIRDER BOX

DETAIL WITH WALL BOX

CENTER POST DETAIL WITH PINTLE

Figure 8·2 Mill construction items and details

pintle key together in such a manner as to prevent movement or misalignment. The typical details of mill construction in Figure 8-2 show pintles, bases, and various fasteners used for (and associated with) this type of construction.

Laminated material Laminated wood members can be used in standard mill construction systems. The uniformity of material throughout laminated members often allows the overall size to be reduced somewhat from the dimensions required for solid timber. Laminated wood is usually more smoothly surfaced, which creates a more attractive appearance. The better finished surfaces make the material more pleasant to work with, which increases the speed with which the structure can be erected.

Mill construction utilizes heavy wood decking for floor and roof construction. Laminated decking can provide smoother finished surfaces, thinner decking for the same span, and a much better appearance as the exposed ceiling material.

DETAILING MILL CONSTRUCTION

The details of joints and methods for mill construction remain essentially the same, whether solid timber or laminated wood members are used. The same types of hardware and fasteners are often used, and the only real difference lies in the makeup of the actual members themselves.

Detailing procedures of mill construction follow the general rules for any type of wood or timber structural system. The framing plan may be drawn as a separate item or combined with the regular floor plan. If the floor plan has a large amount of other information pertaining to processes, equipment, etc., it is probably best to draw the framing plan separately.

Framing plans In preparing the framing (or structural floor) plan, the columns or posts are drawn to scale with solid object lines. Beams, girders, and purlins above the posts are indicated with broken lines representing both sides of the member as it would appear when viewed from above.

Decking is shown only partially to indicate the size and direction of the individual pieces. The lines representing the decking continue over the other beams, etc., which provide its support. Wall plates, ledgers, and similar members are shown in their proper location in the same manner used to represent beams and girders. Dimensioning follows standard procedures.

Sections and details Cross sections and details are most important for mill construction. The decision as to which (and how many) details to draw must be based upon a knowledge of the structural system and a judgment as to what is minimally required to explain the system fully to the builder. A partial plan for mill construction appears in Figure 8-3.

Shop drawings and numbering Shop drawings provide for the fabrication of structural components at the mill. Many of the components of mill construction systems are standard items including some trusses and arch forms. Even special (nonstock) trusses and arches are engineered by the manufacturer around his own equipment and fabricating capabilities. It is not necessary, therefore, always to prepare bona fide shop drawings for this type of construction. It is advisable to prepare, as part of the structural drawings, elevation drawings (face view) of the stock or fabricated components showing desired dimensions, such as span, length, height, pitch, etc.

The various members may be given numbers in the manner discussed earlier in this unit or simply called out by nomenclature and size on the drawing. Certainly, a numbered system accompanied by beam-and-column schedules facilitates faster takeoff of the material for bidding, ordering, shipping, and erection purposes.

REGULAR STRUCTURAL SYSTEMS

The regular framing types of wall-bearing, post-and-beam (beam-and-column), and long-span appear frequently in industrial structures. These systems may be used in their pure state, or the combination of details and methods from two or more may occur. Such a combination might be utilized when a wide, clear, central area was desired, flanked by low, smaller areas on each side. Many aircraft structures require the large central area for the planes and lower, narrower areas for shops, offices, and supportive activities. Such a facility could very well rely upon trusses or aches for the center bays and wall-bearing or beam-and-column framing for the side bays.

Trusses Many types of trusses appear in industrial structures. The Howe, Fink, Pratt, scissors, Warren, and bowstring trusses have all been utilized. The bowstring truss has been used frequently, and is produced in laminated wood for long spans.

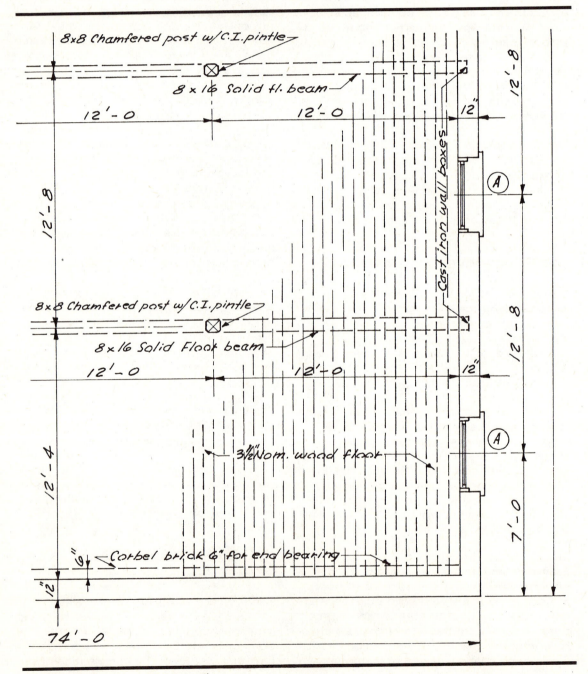

Figure 8·3 Partial plan for mill construction

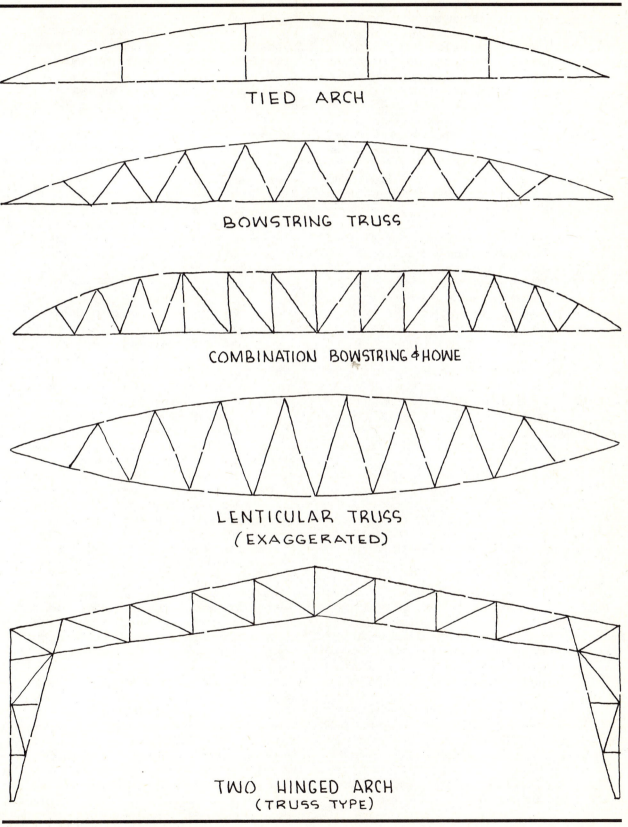

TIED ARCH

BOWSTRING TRUSS

COMBINATION BOWSTRING & HOWE

LENTICULAR TRUSS
(EXAGGERATED)

TWO HINGED ARCH
(TRUSS TYPE)

Figure 8·4 Typical industrial trusses and arches

Arches Arches of numerous shapes provide the structural system for some industrial structures. Some arches create severe stress conditions at the base. The *tied arch* is of the radial arch type, but has the advantage of having the legs tied together before erection. It is actually a self-contained unit and transmits all loads (except wind load) straight downward (vertically) to the supporting wall without the side thrust associated with many other arch forms. The units appearing in Figure 8-4 represent some of the typical trusses and arches of laminated wood produced for the industrial market.

Fastening methods and devices Fastening methods employed for attaching arches and trusses vary with the type of unit and follow two basic classes of details. One type of system works for most units which do not exert side thrust, and the other methods pertain to those units tending to push out at the base. The fastening methods for bowstring trusses and tied arches are, for the most part, identical. The use of heavy steel heel plates is common and provides the necessary anchorage to the supporting wall (Figure 8-5).

The *hinge connection* for radial arches is often used in order to counteract the effects of thermal expansion and contraction, wind load, and side thrust at the base of the arch. Many times the arches are anchored to concrete piers (similar in design to buttresses) which have been integrally poured with the foundation.

The hinge connection, as shown in Figure 8-5, consists of two heavy steel hinge components joined by a high-grade-steel hinge pin. The base component is secured to the concrete pier with heavy steel anchor bolts, while the top hinge unit is attached to the arch with bolts and shear plates.

The vertical legs of other arch types are anchored, in many instances, with steel base units termed *shoes*. These bases are attached to the foundation with embedded anchor bolts and to the arch leg with bolts or lag screws. A typical variety of base connections appears in Figure 8-5.

Details for joining components of heavy solid timber or laminated wood for wall-bearing or post-and-beam industrial systems often resemble those for mill construction. A wide variety of fasteners and fastening devices is available for supporting and joining the various individual members.

Posts (columns) are anchored to the foundation system with devices such as metal shoes, base plates, angle connectors, and strap connectors. The actual hardware is most often fastened to the concrete with anchor bolts, with part of the fastener actually embedded in the concrete in some instances. Posts are attached to the fasteners with lag screws, bolts, or bolts with shear plates or split rings added, depending upon the loading conditions.

The heavy usage of industrial facilities makes it advisable to run the post (or column) foundation to a higher level than the floor (extended above floor level). This practice protects the post from liquids and many impacts from moving equipment. Many post bases enclose only two sides of the post, while other types completely surround the bottom of the post. In detailing bases which surround the post on all four sides, it is good practice to call out vent holes in the base to prevent deterioration of the wood from moisture and decay. Some of the many methods for anchoring wood posts appear in Figure 8-6.

There is a wide range of methods for joining beams to posts, bearing walls, and girders. The attachment may be accomplished with metal hardware, wood bearing blocks, wood tie strips, wood ledgers, or a combination of wood units and metal hardware. The heavy loading conditions present in industrial structures demand fastening methods which result in rigid, solid joints. Some joints encounter such high stresses that bolts will not distribute them enough to prevent failure of the wood components, and split rings or shear plates must be added to increase the bearing surface and decrease the unit stresses to the point where the wood can safely support them. Several joint systems appear in Figure 8-7.

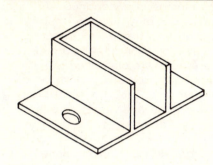

METAL SHOE FOR ARCH BASE

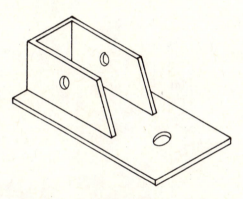

SHOE FOR ARCH BASE

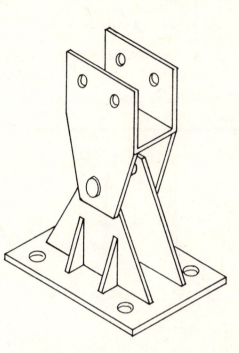

HINGE CONNECTION FOR ARCH

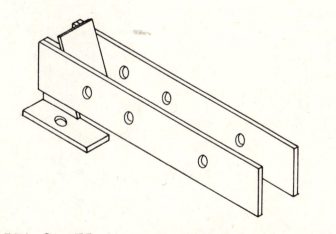

HEEL PLATE CONNECTOR FOR RADIAL TIED ARCH
OR TRUSSES

Figure 8·5 End connectors for arches and trusses

Thick steel straps and plates are a quick means of attaching beams and girders to timber posts. Some joints must be furnished with lateral ties to provide structural continuity between the units. These ties may be in the form of additional wood components, or they may be standard iron or steel bars with the ends bent downward (sometimes called *dogs*).

Multistory structures often demand that

posts or columns run continuously from foundation to roof. In such instances, beams and girders frame into the sides of the posts rather than resting on top of them. When posts are placed on top of one another for such multistory work, a standard practice is to place a steel bearing plate between the post ends to ensure a flat bearing surface.

Beam-to-girder and purlin-to-beam attach-

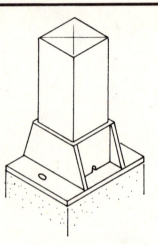

ONE-PIECE BASE WITH VENTS

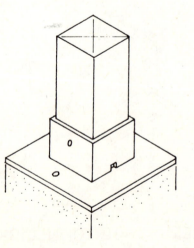

BEARING PLATE WITH STRAPS

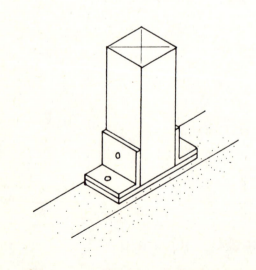

BEARING PLATE WITH ANGLES

ONE-PIECE VENTED BASE

Figure 8·6 Typical bases for wood posts

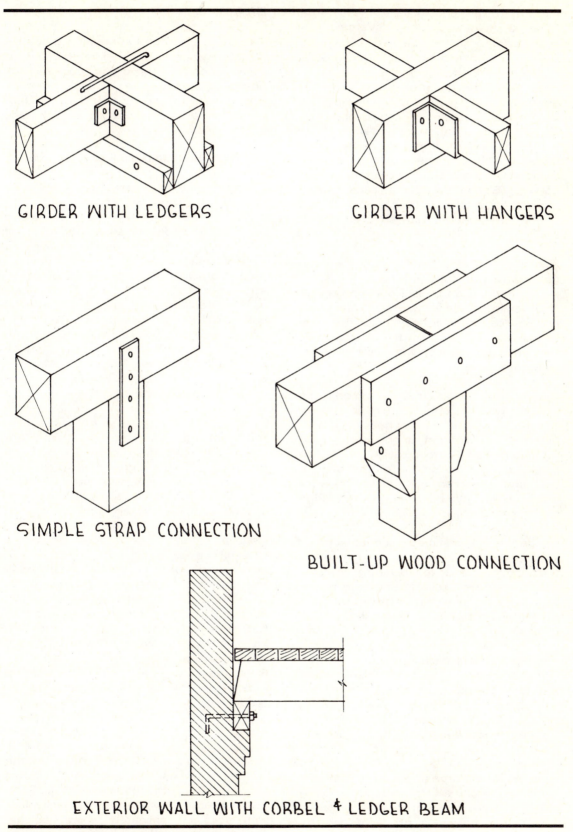

GIRDER WITH LEDGERS

GIRDER WITH HANGERS

SIMPLE STRAP CONNECTION

BUILT-UP WOOD CONNECTION

EXTERIOR WALL WITH CORBEL & LEDGER BEAM

Figure 8·7 Typical heavy timber connections

ment is often accomplished with steel hangers or straps. Some systems rely upon a *ledger* to provide end-bearing for beams or purlins. The ledger is an additional wood member which is attached to the side of the beam or girder and runs continuously along its length. The ledger is not as deep as the main member and is attached so that its bottom surface is flush with the bottom surface of the main member. This practice results in a shelf upon which the ends of beams or purlins can rest. Since the ledger is called upon to support large total loads, it is often bolted to the main member with shear plates or split rings (see Figure 8-7).

Beams can be attached to bearing walls in many different fashions. Pilasters may be constructed as an integral part of the wall. The height of the pilasters may be held enough below wall height so that the beams secured on top of the pilaster have their top surface flush with the top of the wall plate.

Recesses or pockets may be formed in the wall (as with mill construction) to receive the ends of girders or beams. Heavy members should rest upon bearing surfaces provided by steel plates. Beams can be placed directly on top of the bearing wall and anchored with clip angles or straps which have been securely attached to the masonry work.

Ledgers are sometimes used to support beam ends at bearing walls. In such cases, the ledger is a large structural member which has been securely attached to the wall. The attachment may be by through-bolting and may necessitate steel anchor straps tying it further to the wall.

Fire walls create certain problems in providing end-bearing for beams. Great care should be taken to preserve the fire-stopping characteristics of such a wall. *Corbeling* the brickwork to provide beam-bearing is a frequent practice as detailed in Figure 8-7. Corbeling consists of permitting successive courses of brick increasingly to overhang (extend past) the main wall line. When this is done with several consecutive courses, it creates a brick shelf or ledge upon which the ends of beams or the bottom of ledgers can rest.

BRACING

The entire structural system for industrial structures, as mentioned in the unit on steel, encounters numerous shocks and vibrations from the type of usage it serves. Such systems should therefore be provided with adequate bracing to withstand such conditions.

Diagonal and longitudinal bracing provide a means of tying the entire structural system together, and furnish a damping capacity to absorb the impacts and vibrations created by heavy equipment within the building. Bracing may take the form of steel rods, additional wood members, or even stressed cables. The addition of proper bracing is extremely important in structures where moving equipment, such as overhead traveling cranes, is supported or braced from the structural system itself. Longitudinal struts are most important in providing a continuity and structural tie between roof trusses. The struts are often placed at the ridge, eaves, and in one or more positions between the bottom chords.

Wind bracing Industrial structures are often extremely large buildings and are often subjected to high wind-load factors. Wind bracing for any structure is a good (and common) practice and becomes most important in industrial applications.

Regular winds (those directed against a single face of a building) produce high amounts of load. Diagonal winds (those directed at the corner of a structure) produce the same high loading factors and, in addition, tend to rack or twist the structure. Bracing at corners and in alternate bays serves to prevent wall damage to the structure from direct and diagonal wind. Roof trusses are particularly prone to the racking and twisting action produced by wind.

Diagonal braces between trusses can counteract the wind-induced racking. These braces should be installed in the same planes as the upper and bottom chords. Much of the tendency of trusses to twist is reduced when the roof is decked with solid or laminated wood decking. These thick wood units (often single or double tongue-and-groove) create a mono-lithic deck which adds appreciably to the overall stiffness and rigidity of the entire roof structural system. When compared to joists with plywood or diagonal sheathing, the heavy deck has a decided advantage in insulating qualities and fire rating.

The degree of stiffness or rigidity which can be achieved with any structure varies with the type of structural system em-ployed. The beams or trusses attached directly to heavy load-bearing masonry walls more easily provide a high degree of rigidity than do beams and trusses sup-

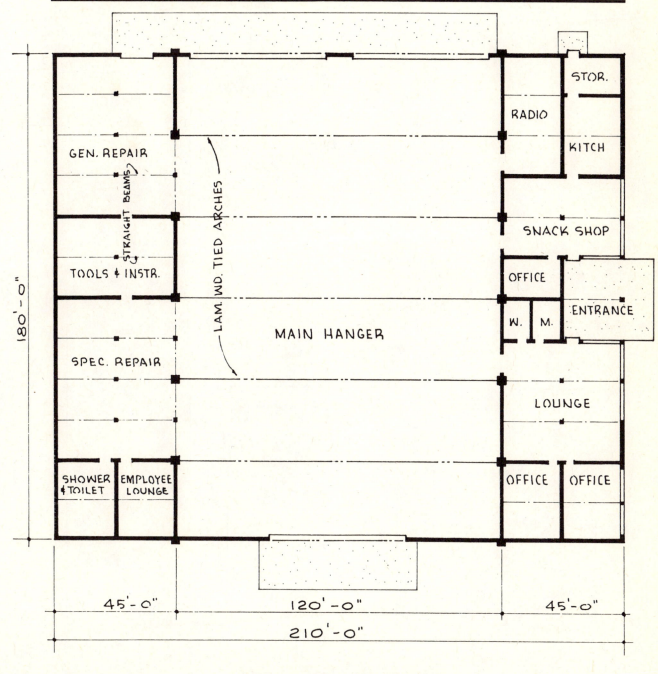

PRELIM. FLOOR PLAN

Figure 8·8 Small airport structure

ported by posts or columns. Much care must be taken in designing joints for the latter methods, to assure that the members will not move under actual load conditions.

EXAMPLE OF COMBINATION FRAMING SYSTEM

The triple-bay structure shown in Figure 8-8 is an example of a combination system

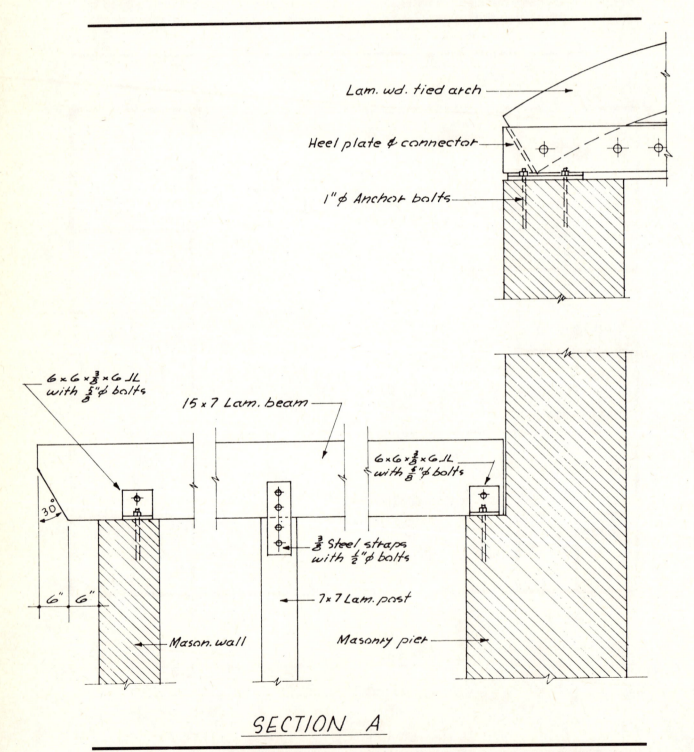

SECTION A

Figure 8·9 Sectional detail – airport building

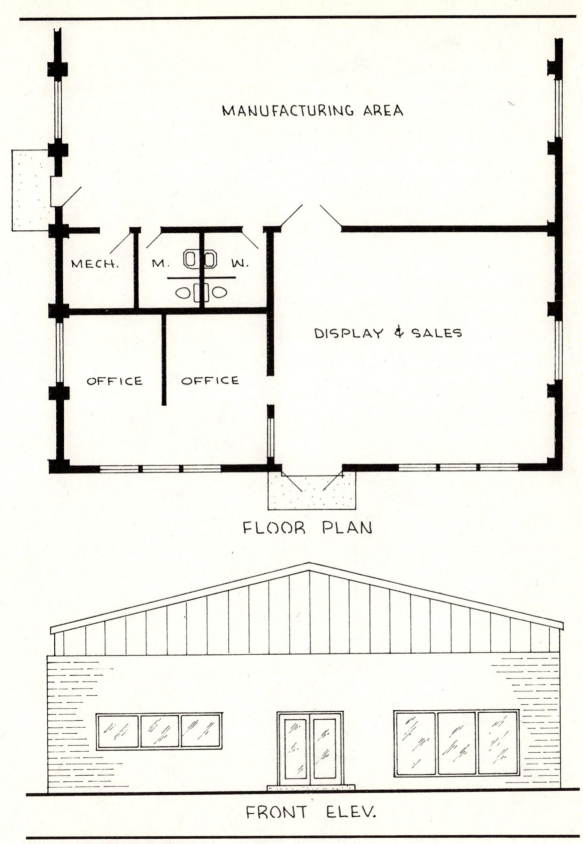

MANUFACTURING AREA

MECH.

M. W.

DISPLAY & SALES

OFFICE OFFICE

FLOOR PLAN

FRONT ELEV.

Figure 8·10 Small factory building

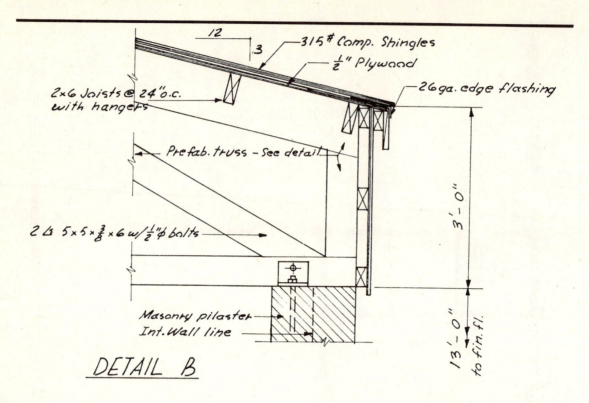

12
3

315# Comp. Shingles

½" Plywood

26 ga. edge flashing

2×6 Joists @ 24" o.c.
with hangers

Prefab. truss – See detail

2 ⌐ 5×5×⅜×6 w/½"φ bolts

3'–0"

Masonry pilaster
Int. Wall line

13'–0"
to fin. fl.

DETAIL B

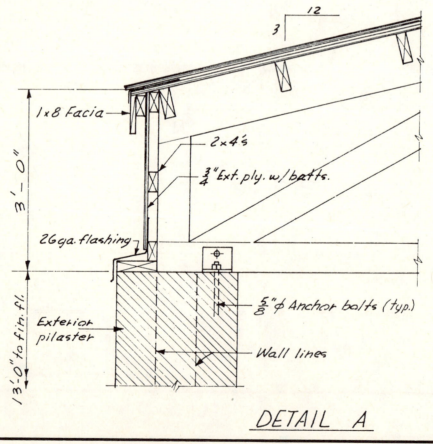

12
3

1×8 Facia

2×4's

¾" Ext. ply. w/ batts.

3'–0"

26 ga. flashing

⅝"φ Anchor bolts (typ.)

13'–0" to fin. fl.

Exterior
pilaster

Wall lines

DETAIL A

Figure 8-11   Truss-roof factory building details

incorporating long-span, post-and-beam, and wall-bearing structural systems. The building represented is for a small airport and combines aircraft service, maintenance, storage, and passenger service areas. The main portion of the structure (aircraft storage and service) is spanned by laminated tied arches. These arches provide clear space, and the interior of the main area is usable without the interference of posts.

The side bays are framed with laminated beams joined to the piers and supported by wood posts on the interior and by load-bearing masonry walls on the exterior. The beams and arches are topped with 4-in. solid wood decking with double tongue-and-groove edges. The lower roofed side bays house the parts and service departments, storage, office space, cafe, lounge, ticket counters, and passenger service areas.

In the plan in Figure 8-8, the columns (posts) and walls are drawn with solid object lines, and the beams and arches are indicated with broken lines. Notice that the piers for the arches are in the building interior and are indicated on the plan.

Sections and details are necessary to provide an accurate picture of what the struc-

tural system is to entail. The drawings in Figure 8-9 indicate some of the methods of joining and attaching the various components and the hardware necessary for the connections.

EXAMPLE OF TRUSSED CONSTRUCTION

The industrial building shown in the partial plan and elevation in Figure 8-10 is designed for a small manufactuer of farm equipment. The structure depends upon load-bearing concrete block walls with pilasters and laminated wood trusses.

The building provides space for production, storage, and offices. Notice that the pilasters along one side of the building have been doubled in size and project from both the interior and exterior wall surfaces. This was done to permit ease in adding space utilizing similar trusses. The exterior pilasters will provide end-bearing for the trusses of the next addition to the building.

The end conditions for the trusses are of two types due to the exterior pilasters on one side. Two separate details are necessary to explain what is intended. The details in Figure 8-11 provide additional information for the contractor.

SUMMARY

Wood has been an important material for many industrial structures in the past and continues in use today. Like any of the other structural materials, wood does not fit every construction need. The decision to use the material should be well thought out to determine that the choice will result in a safe, functional structure.

Mill construction has been standardized through the years and has certain attributes which set it apart from other construction types. This is particularly true of the details of the various joints and connections.

Industrial structures currently make use of the three main system types of long-span, post-and-beam, and wall-bearing framing. In addition, many installations lend themselves to a combination of the three types.

Drawing and dimensioning practices for wood industrial structures follow closely the conventions used in drawing and detailing steel structures. As with steel, the detailer must provide clear, concise information to facilitate fast, accurate estimating and efficient fabrication and erection.

The structural system for industrial structures is subject to many shocks and vibrations not encountered in residential and commercial work. This factor must be considered when detailing the system. The size of such structures amplifies the problems of wind load and twisting or racking of the structure.

A well-detailed wood structural system, when applied to a building for which it is well suited, can result in an efficient, sound, and economical structure to house the operations of industry.

Problems—Chapter 8

1. Prepare typical mill construction details for the beam and post shown.

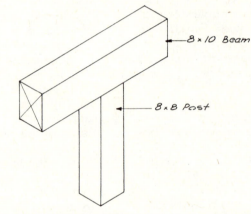

— 8 × 10 Beam

— 8 × 8 Post

2. The beam in the illustration is to be supported at the end by the 12-in. brick wall. Draw the end details in standard mill construction form.

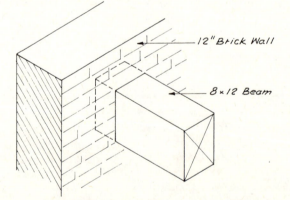

— 12" Brick Wall

— 8 × 12 Beam

3. The column (post) in the figure runs for two stories. Detail all joint conditions for the column-to-beam joints in the mill construction method. (see the diagram on the next page)

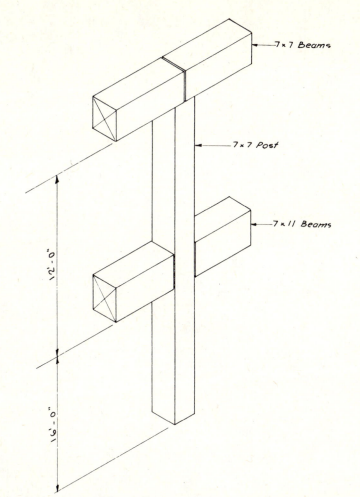

4. A wood post (column) is 8 × 8 in. square. Detail the base anchoring system four times, using each of the following devices: (a) metal shoe, (b) base plate, (c) angle connectors, and (d) strap connectors.

5. Two 6 × 8 in. beams are to be supported by an 8 × 14 in. girder. Detail the joint, using three different methods of attachment.

6. The illustration gives the information for a portion of an industrial manufacturing plant. Prepare a partial structural plan for the building, and provide the necessary details for the construction, based upon the following:
   a. All girders are 8 × 20 in.
   b. All beams are 6 × 12 in.
   c. All posts are 8 × 8 in.
   d. Exterior walls are 12-in. brick.
   e. Roof decking is 4-in. double tongue-and-groove wood deck.
   f. Height from finished floor to bottom of deck is 16 ft 0 in.
   g. The structure has a flat roof.
   h. Use typical steel connectors with ⅝-in. bolts

   (see the diagram on the next page)

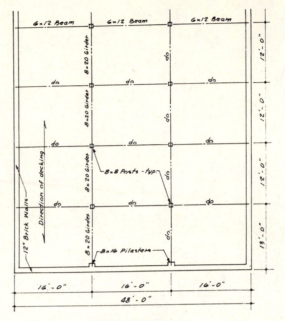

PARTIAL FLOOR PLAN

7. The small industrial building has 12-in. concrete block walls and utilizes wood scissors trusses. Obtain manufacturers' literature for the type of truss shown, and prepare plan, details, and truss details for the building. The wall height is to be 10 ft 0 in.

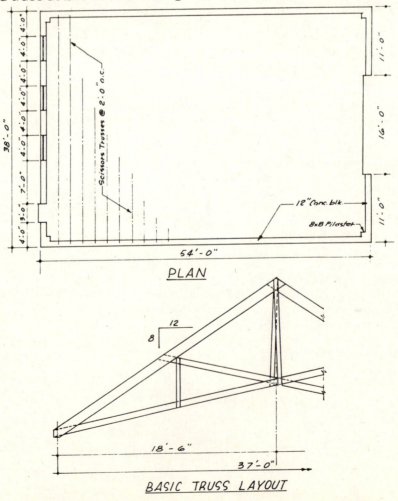

PLAN

BASIC TRUSS LAYOUT

unit three

Courtesy of United Nations

concrete

After Studying This Chapter You Should Be Able To:

1. Describe the composition of the earliest concrete, and cite the area and approximate date of its first use.
2. Trace the development of the use of concrete, and give the name and location of the man credited with creating portland cement.
3. Explain how concrete construction changed during the period of World War I.
4. Explain how the use of concrete structural systems has changed since World War II.
5. Define, in clear terms, what concrete is, and cite its components.
6. Describe the composition of modern cements and the process used in producing them.
7. Describe the reaction called *hydration*, and explain why it is necessary in concrete work.

16. Explain how prestressed concrete varies from regular precast concrete, and cite the advantages it offers.
17. List the various manufactured units available in precast and prestressed form.
18. List, and describe in detail, the two principal sets of drawings prepared for structural concrete systems.

INTRODUCTION

Concrete, in some form, has been used for construction purposes for almost 2,000 years. The early uses were certainly limited and did not begin to develop the potentials of the material as we know it today.

Possibly the first use of concrete as a legitimate structural material was in the structures of the Romans around A.D. 300. Early Roman builders prepared a concrete based upon a combination of sand, stones

---

chapter 9
the material-
concrete

---

8. List factors which affect the quality of concrete.
9. Explain what determines the classifications of *fine* and *coarse* aggregates.
10. Explain why it is necessary to consolidate concrete in forms, and explain how the process is achieved.
11. List the variables which affect the strength of concrete.
12. Describe tests used to determine the adequacy of concrete during and after its placement.
13. Explain how reinforcing adds to the adequacy of concrete structural members.
14. Describe the various types of reinforcing used in concrete construction.
15. Define the two main categories of cast-in-place and precast concrete construction.

or brick fragments, water, and cement. Their cement, called *Pozzolana cement*, was made of powdered volcanic material indigenous to their geographic location.

Early Roman concrete was placed in much the same manner used today. The builders constructed wood forms and poured the concrete into the area between the forms. After the concrete had set, the forms were stripped (removed). Many Roman buildings were actually concrete structures veneered (surfaced) with marble, mosaic, or painted plaster. Figure 9-1 indicates some early formwork used by Roman builders.

Much of the similarity between various Roman structures throughout the Roman Empire was due, in part, to the influence which concrete had upon structural forms. The Romans developed many rather advanced methods for using concrete. Roman domes are indicative of the degree to which these early builders developed the technology of concrete construction. Domes and vaults were often formed with recesses (or coffers) on the underside, thereby reducing the overall weight of the structural system. The same practice is used currently in pouring many types of roof and floor slabs.

The influence of the Roman Empire slowly spread the use of structural concrete to many other locations. The early 1700s found concrete in use in some form throughout most of the world. Most early

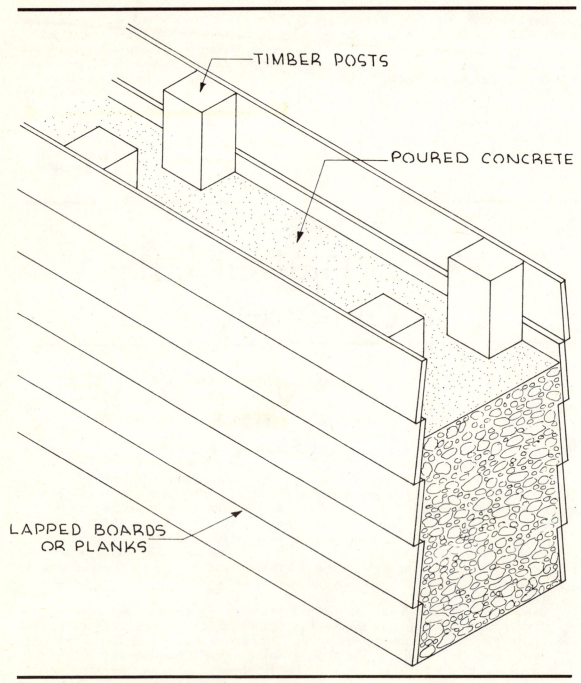

Figure 9·1 One type of early Roman formwork

uses were rather cumbersome, and efforts to provide reinforcing were crude and often inefficient.

The 1800s saw concrete used with reinforcing in an adequate manner. Shortly after the development of iron and steel structural systems, reinforced concrete gained markedly in popularity. One of the factors contributing to the increased acceptance of concrete was the development of the first *portland cement.*

Joseph Aspdin, an English mason, obtained a patent for his portland cement in 1824. The concrete produced with Aspdin's cement closely resembled a type of natural limestone produced by the quarries on the English Isle of Portland. The new product was so named because of its resemblance to the Portland limestone.

Portland cement was used in England and its neighboring European countries for some 50 years prior to its manufacture on the North American continent. The first recorded production of portland cement in North America was initiated with a cement plant in Coplay, Pennsylvania, in 1872. The first similar plant in Canada was established in Hull, Quebec, where production began in 1889.

Complete structural systems of reinforced concrete designed in a standardized manner are often credited, in their development, to French designer François Hennebique. The efforts of other designers had been mostly experimental and sporadic prior to Hennebique's designs. Individual early examples, such as Max Berg's 213-ft-diameter concrete dome in 1870, had been successful one-of-a-kind undertakings, but Hennebique's work in 1892 gave reinforced concrete a firmer base of standardization, with more predictable standards for replication.

Improvement of reinforced concrete design and construction methods advanced at a steady pace in France and Germany prior to World War I, with extreme degrees of increased use and adequacy of design during the war. The period of increased activities in England and the United States occurred in the latter part of the 1920s.

The shapes of structures changed with the increase in the knowledge of reinforced concrete design, as evidenced by Eugene Freyssinet's airship hangar built at Orly airfield in France in 1916. This huge concrete structure was provided with clear interior space through the use of reinforced concrete parabolic vaults.

Many of the early technical developments resulting in complete concrete structural systems were patented by the designer. These systems were marketed in a proprietary manner. Most of the early applications were directed toward industrial structures and bridges. The innovative industrial and bridge designs of the Swiss engineer Maillart spread the use of reinforced concrete systems to Italy and South America.

The period between 1920 and World War II saw daring uses of concrete systems, as evidenced in the works of Frank Lloyd Wright, Eric Mendelsohn, Eero Saarinen, Pietro Belluschi, and the Perret brothers.

Great strides have been made since World War II in the development of reinforced, precast, and prestressed concrete structural systems. No longer have designers limited structural concrete to industrial and bridge construction. All categories of buildings have been constructed utilizing various concrete systems, including public buildings, educational facilities, stadiums, churches, commercial and office structures, and residences. Many renowned architects and engineers of the current century have created exciting new structural forms in reinforced and prestressed concrete. Designers P. L. Nervi, Alvar Aalto, Edward Torroja, and Felix Candella have expanded the potential of concrete immensely. Innovation and experimentation continue to expand the many potential uses and forms of structural concrete.

COMPOSITION OF CONCRETE

Concrete is a building material composed of sand and gravel (aggregate) bonded together with cement. The cement, combined with water during the mixing process, forms a paste which coats the aggregate particles and fills the space between them. Once the concrete is in place, it is allowed to dry and hardens to a stonelike solid.

Cement paste The cement paste makes up approximately 25 to 40 percent of the volume of the concrete, with the balance of the volume constituted by the aggregate. The variance in volume percentages occurs with the different proportions used for lean and rich mixes (referring to the proportions of cement used). The cement itself is a powdered mixture composed essentially of burned lime and clay. The same cement can be used for mortar (for laying brick and block) or for concrete. The difference between mortar and concrete is that mortar contains finely graded sand rather than the sand and coarse aggregates used in concrete.

Portland cement One of the most frequently used cements is called portland cement. The modern version of this mixture differs somewhat from its predecessor of 1824 and is a rigidly controlled combination of several basic materials. A modern portland cement typically contains limestone, cement rock, oyster and marl shells, clay, silica sand, shale, and iron ore. These basic ingredients are pulverized and heated in a kiln. The product emerging from the kiln is portland cement clinker. The clinker is cooled, pulverized, and combined with pulverized gypsum to form the finished cement.

Gypsum The gypsum which is added to the clinker serves in regulating the setting time of the cement. Concrete can be made to solidify more slowly or more quickly by varying the amount of gypsum contained in the cement.

Marketing cement Portland cement is marketed in bulk and in individual bags. The bulk cement is shipped in special trucks or rail cars. Each bag produced in the United States contains 94 lb of cement packaged in a heavy paper bag. The cement is ground extremely fine, a factor very necessary to its effectiveness. Current processes used in producing portland cement result in a product so finely ground that nearly all of it will pass through a sieve which has 40,000 openings per square inch.

Chemical reactions Cements which react with water for setting and hardening are called *hydraulic* cements. Portland cement depends upon such a chemical reaction to set and harden. The chemical reaction itself is referred to as *hydration*.

QUALITY OF CONCRETE

The quality of concrete depends upon a number of factors. Basic to the quality of any concrete mix is the quality of the *paste* (cement and water). This factor varies with the ratio of water to cement. As the paste is thinned more with water, the strength decreases, and the resulting concrete is much more prone to damage from weather.

Aggregates The choice of aggregates is extremely important and affects the strength and the weather resistance of the concrete. Fine aggregates generally consist of natural or manufactured sand with particle sizes less than 1/4 in. Coarse aggregates are classified as those having particle sizes greater than 1/4 in. Both types of aggregates are used in structural concrete. Some sources provide aggregate containing materials which have an injurious effect on the concrete. Aggregate with high iron content produces rust spots in concrete, which gives it an unsatisfactory appearance. Aggregate containing organic impurities, silt, clay, coal, lignite, and fibrous materials can adversely affect the strength and durability of concrete.

Water The water used in mixing concrete can affect the appearance, strength, and setting time of the mixture. Water containing high amounts of dissolved solids and suspended solids can seriously alter the quality of concrete. Water containing significant proportions of various kinds of salts, acids, alkane, sugar, oils, and algae should not be used for concrete mixtures.

Placing The placement and curing procedures of concrete have a decided effect on its quality. Concrete must be placed properly in the forms, and it is often necessary to consolidate the mixture by the use of spades, puddling sticks, tampers, or vibrators. The consolidation process assures that all voids have been filled with concrete and that the various layers of individual pours have been sufficiently mixed together.

Temperature and curing The quality of concrete can be adversely affected when temperatures during the placement process are extremely hot or extremely cold. Special precautions can be taken in such instances, and special admixtures are available designed for hot- or cold-weather work. After placement, the quality of concrete can be maintained by careful curing. The curing process guards against the loss of necessary moisture during the early hardening period. It also helps maintain the most favorable temperature for the hydration reaction. Several methods are used for curing. Flooding new slabs with water, sprinkling slabs, or covering them with wet coverings such as burlap are of advantage in hot weather. Waterproof coverings also serve to maintain favorable moisture and temperature conditions within the new concrete. Waterproof curing papers and plastic sheets are widely used for this purpose. Manufactured concrete units are often cured with the application of steam, either at atmospheric pressure or in an autoclave at high pressure.

STRENGTH OF CONCRETE

The strength of concrete depends upon a number of variables. The proportions of cement to water and aggregate have a direct bearing on the strength of the concrete. The choice of aggregates can affect the strength, as can the ratio of fine to coarse aggregates used in the mixture. The actual strength of structural concrete units is, of course, increased with the addition of proper reinforcing steel. Prestressing of concrete units provides additional assistance in developing high allowable stresses.

Tests of concrete Testing of sample mixtures prior to actual construction, and various tests performed during and after the actual placement of concrete, help assure that adequate concrete strengths are

being achieved. Representative samples of concrete are taken from various batches of concrete during the placement process. These samples are used in two different manners. A *slump* test is performed with part of the samples. This consists of filling a standardized metal cone with the wet concrete. The cone is then inverted and removed. The wet mixture will sag or slump to less than its height when contained in the metal cone. A measurement of the loss of height of the unsupported concrete (the actual amount of slump) gives an indication of the consistency of the concrete.

The slump test provides a means of assuring that the various batches of concrete are of uniform consistency. It helps to avoid concrete which is too wet or too stiff. Guidelines have been drawn by various agencies specifying maximum and minimum acceptable slump values for the various construction applications, such as slabs, footings, columns, beams, etc.

Other concrete samples are taken during the placement process and are placed into small cylindrical forms. These samples are allowed to harden in the forms and result in concrete test cylinders 6 in. in diameter and 12 in. tall. The cylinders are subjected to tests at various intervals, generally ranging from 7 to 28 days. The cylinders are crushed on a testing machine, and the compressive strengths of each series of samples are recorded. The strength of concrete varies significantly during the first month after placement, and the tests provide a method of determining that the strength of concrete in the project is adequate for the intended design.

CONCRETE REINFORCING

The stability, rigidity, and resistance to cracking of concrete structural units and members can be increased by embedding steel reinforcing bars, cables, or mesh in the concrete as it is being placed. Bars, cables, and mesh may be used separately, or a member or slab may have a combination of two or all three types. Practically all concrete used in structural applications is reinforced in some manner.

Bars Reinforcing bars may be round or square and may have smooth surfaces or surfaces deformed to provide better *bonding* (adherence) of the concrete to the bar. The round deformed bar is the most used and is produced in a wide range of diameters varying in increments of 1/8 in. The standard method of calling out round bars is to give the number of 1/8s contained in the diameter. A 5/8-in.-diameter bar is shown as a #5 bar, a 3/4 in. as a #6, etc.

Mesh Concrete slabs are often provided with a reinforcing grid of steel wires. This grid consists of steel wires, running in two directions (perpendicular to each other), which have been welded at the points where they cross. The grid is referred to as *welded wire fabric* or *welded wire mesh* and abbreviated as W.W.F. or W.W.M.

The actual size of welded wire fabric is given in terms of the size of the square or rectangle formed by the wires, and the gage (size) of the steel wires running in each direction. The conventional system cites the spacing size first, followed by the wire gages. A reinforcing fabric with 6-in. squares and 10-gage wires running in both directions appears on the drawing as 6 × 6 / 10 × 10 W.W.F. (or W.W.M.).

METHODS OF USING CONCRETE

There are two main categories which encompass all concrete structural methods. Concrete is either *cast in place* on the actual job site in its permanent location, or it is *precast* and erected after it has been cured. *Prestressed* concrete often combines precasting with prestressing.

Cast in place Cast-in-place concrete is probably most used and dates from the methods used by the early Romans. Forms are constructed in place, and the member or slab is cast in the forms. After hardening, the forms are removed. Footings, foundations, and earth-supported floor slabs are frequently poured in this fashion. Upper floor slabs and walls, beams and columns are frequently cast in place by constructing forms supported by temporary shoring (bracing).

Precast Precast concrete is widely used and consists of casting the concrete units in a more controlled environment such as a shop, curing them, removing them from the forms, and transporting them to the site for erection. Since the forms for precasting can be reused, it is economically feasible to construct higher-quality forms. The control provided in precasting can provide more consistent concrete mixtures and curing processes. These factors tend to result in better dimensional control and in much better surface finishes.

Large precast units are relatively heavy, and provisions must be made for moving and erecting them. Hardware must also be provided for attaching the units to the structural system. Steel lifting rings are often fastened to the internal reinforcing steel for the purpose of attaching crane cables for lifting, loading, unloading, and erection. Angle connectors are sometimes welded to the reinforcing steel, which, in turn, can be bolted or welded to the structural system. Reinforcing is also permitted to extend past the surface of the units, so that the reinforcing of adjacent units can be welded together to join them.

Prestressed The elastic properties and allowable stresses of concrete are increased appreciably by prestressing the concrete. Prestressing frequently permits a decrease in the size of unit required for a given load condition or span. Most prestressing is applied to precast units, but prestressing can be accomplished in the field for cast-in-place units as well.

Wires or cables are used for reinforcing prestressed units. These wires or cables are stretched before the concrete is placed around them. After the concrete has hardened, the tension devices are released, and the wires or cables attempt to resume their original shape and length.

The pressures set up in the concrete bonded to the wires after releasing them from their stretched position create compression within the concrete itself. This internal compression prevents cracking and sagging of the prestressed units. Since prestressing overcomes some of the units' tendency to sag, the overall depth of beam members can often be noticeably decreased from that required for regular reinforced concrete members.

Most prestressing is done under shop conditions, and rigid controls are applied. The quality of prestressed units is most important to both their structural performance and their appearance. Since forms are carefully constructed and maintained, most prestressed units have excellent surface finishes.

PRECAST CONCRETE PRODUCTS

A wide range of concrete products is produced in precast form. The much-used concrete block is a precast unit and is available in both lightweight and structural forms. Blocks are carefully cast and cured, and dimensions run fairly uniform for any given type of block. These units are used for laying walls and are available in a number of thicknesses. The most-used concrete blocks are nominally 4 × 8 × 16 in., 6 × 8 × 16 in., 8 × 8 × 16 in., and 12 × 8 × 16 in. The actual dimensions are smaller than the nominal dimensions so that mortar joints can be included.

Concrete roof deck panels and wall panels are precast by some plants. These units contain reinforcing, but many are not prestressed. The usual range of thickness for such panels is from 2 to 4 in. Lightweight aggregates may be used for precast panels, resulting in less total weight and providing the extra benefit of allowing roofing materials to be nailed directly to the concrete. Roof panels are available in flat and channel slab forms.

A number of surface finishes are available in precast panels. The exposed aggregate finishes provide a heavily textured surface and the opportunity (through specifying various colored aggregates) of obtaining colored panels. The exposed aggregate is actually covered by the concrete when poured. After the panel has hardened sufficiently, and while the surface is still fairly soft, the surface is hosed with water under pressure or brushed with a stiff broom. This process removes the cement paste on the top of the aggregate and exposes it to view. The aggregate remains attached because it is bonded to the panel by the cement paste behind it.

Beams and columns are produced by precasting and have excellent dimensional tolerances. The surface finishes produced by precasting make the units suitable for exposed applications. The prestressed units have larger allowable spans and loads, but regular precast units can function quite well when designed adequately for the particular structural system.

PRESTRESSED CONCRETE PRODUCTS

The practice of prestressing adds considerably to the capabilities of precast concrete units. Many structural components are available in prestressed and precast form, as shown in Figure 9-2. Wall panels and hollow-core slabs are frequently used. The hollow-core slabs are used for floor and roof slabs and can generally span up to 40 ft.

Prestressed columns are produced in square and octagonal form in sizes beginning with 6 in. Concrete piles for supporting foundations in unsatisfactory soil conditions are produced in similar shapes and are furnished 10 in. square and larger. The ends of columns and piles are available in various configurations to fit the type of structural system being used.

Girders are produced as standard prestressed items and are generally of the *I-girder* and *box-girder* types. Heavy load conditions may be overcome by using the hollow box girders as a roof or floor deck with grouting between each member. The I-girder is applicable to spans up to 120 ft and often serves as the principal girder in beam-and-decking systems. The prestressed beam unit often used with double and single T-units and hollow-core slabs is the *inverted T-beam*. This unit provides heavy flanges to receive edges and ends of slabs and secondary beams.

Standard prestressed units include four general types of units which combine structural members (beams or joists) with

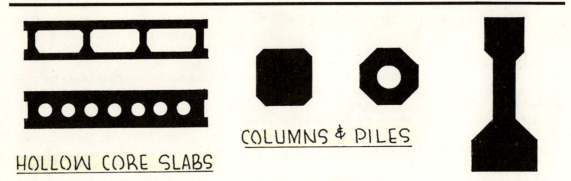

HOLLOW CORE SLABS

COLUMNS & PILES

"I" GIRDER

BOX GIRDER

CHANNEL SLAB

MONOWING ("F") SECTION

INVERTED "T" BEAM

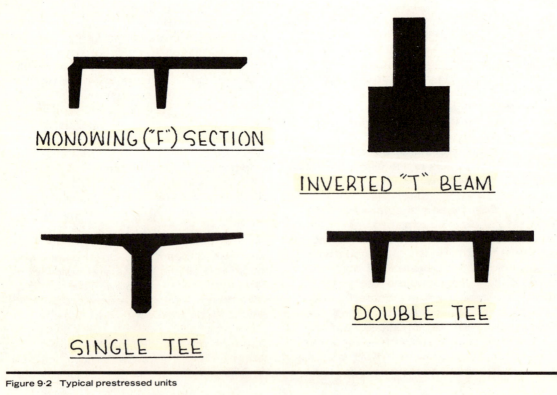

SINGLE TEE

DOUBLE TEE

Figure 9·2 Typical prestressed units

roof or floor deck panels. The *channel slab* is an extremely rigid unit and is used in applications involving heavy floor and roof loads. The *monowing* or *F-section* provides roof and floor deck members and spans up to 60 ft. The deck is cast with two beams per unit. This unit results in hidden joints and offers an excellent exposed appearance.

Double and single T-members combine decking with two beams or a single beam, and permit spans up to 125 ft. The width of such units usually ranges from 6 to 10 ft. With deck and beams combined, these units permit extremely fast erection and enclosure. Appearance of the T's is similar to that of the monowing, except that greater care must be taken in finishing the exposed joints between units.

It is a frequent practice to pour a thin topping slab over prestressed floor systems. Such a slab adds to the continuity of the system and provides the thickness necessary for concealing electrical conduit. The topping slab results in a better surface for the application of floor tile or carpet. The topping slab is provided with wire reinforcing mesh to prevent cracking of the slab and to provide continuity of the slab over the joints of the prestressed units. For some applications, the topping slab is of lightweight or insulating concrete. Insulating slabs are of particular benefit as topping over prestressed roof units.

DRAWINGS NECESSARY FOR CONCRETE CONSTRUCTION

The basic types of drawings necessary to detail poured-in-place concrete structural systems consist of two principal sets called *engineering drawings* and *placing (or placement) drawings*.

Engineering drawings are prepared in the design office and show the design of the structure, including dimensioned location of members, size of individual members, reinforcing of members, and related information. These drawings are similar to the design drawings prepared for steel structures.

Placing drawings are actually detail drawings which show shape, size, and location of the reinforcing bars in the structural system. Information is provided on the method of placement, and schedules are given for beams, columns, girders, joints, etc.

The complete set of concrete drawings includes plans, elevations, sections, details, and schedules for the various components. Scales vary, but plans are most often drawn at scales of ⅛ or ¼ in. = 1 ft 0 in. Elevations are drawn at ¼, ⅜, or ½ in. = 1 ft 0 in. Sections range in scale from ¼ to 1 in. = 1 ft 0 in. Details are enlarged as necessary and often range from ¾ to 3 in. = 1 ft 0 in.

STANDARD LINES

As with any type of structural drawing, the drawings for structural concrete are based on a standardized series of lines. Each type of line conveys a specific piece of information. Figure 9-3 contains the various line types used in the work. Edges of such components as footings and other unexposed edges are represented by broken lines (actually hidden lines), while the edges of exposed concrete elements are shown with heavy solid object lines.

Center lines are the same for concrete work as they are for the other types of structural systems. Since reinforcement is extremely important in concrete work, an extremely heavy solid line is used to represent the reinforcing bars. To avoid confusion with object and reinforcing lines, dimension and extension lines are solid lighter lines of equal weight to the center lines.

SYMBOLS AND ABBREVIATIONS

Standard abbreviations and symbols have evolved for use in detailing concrete structural systems. The symbols and abbreviations shown in Figure 9-4 are found throughout the industry. Notice that the bar size (or number) is preceded by the symbol # indicating number. Round bars are designated with the circle and inclined

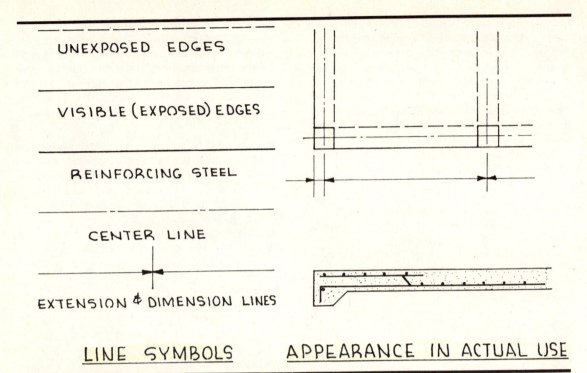

UNEXPOSED EDGES

VISIBLE (EXPOSED) EDGES

REINFORCING STEEL

CENTER LINE

EXTENSION & DIMENSION LINES

LINE SYMBOLS   APPEARANCE IN ACTUAL USE

Figure 9-3 Lines used to detail concrete

slash combined, as ∅. Square bars are indicated with a small square, without the slash used for circular stock. Spacing of bars is given from center to center, and the @ symbol is used in the form of 3 @ 6 in. The example would indicate #3 bars spaced so that there is a distance of 6 in. between their centers.

It is important to indicate the exact area in which given bars are to be used. The line with arrowheads on each end is employed to define the area in such cases. Equally important is the direction in which the bars are to run. The line and arrowheads shown in Figure 9-4 define the direction of the bars in a given area.

A number of abbreviations are used to provide additional information. Bars may be called out as plain P, bent Bt, and straight St. Stirrups are called out as Stir, spirals as Sp, and column ties as CT.

A whole series of abbreviations points out location of reinforcing in relation to the concrete unit itself. I.F. stands for the inside face, F.F. for far face, and E.F. for each face. The top is designated with the capital

T, and Bot. indicates bottom. Since reinforcing often runs in two directions (in the form of a grid), the abbreviation E.W. stands for each way.

The symbols and abbreviations shown are used both on actual drawings and in schedules and notes.

MARKS

The various structural components for concrete systems are labeled or marked to indicate their function and location. The marking system designates the floor or tier level, the type of member, and its location in the system. The third beam in a system, occurring at the fifth-floor level, would be labeled 5 B 3. The first numeral establishes floor level, the letter indicates that the member is a beam, and the last numeral cites the number of the member as it exists in the system of beams at the particular floor level.

The following capital letters are used to mark the various components:

#	Precedes bar size	@	Precedes center-to-center spacing
φ	Designates round bars		Designates bar direction
□	Designates square bars		Defines area included by bars

Pl	Plain bar		I.F.	Inside face
Bt	Bent		O.F.	Outside face
St	Straight		N.F.	Near face
Stir	Stirrup		F.F.	Far face
Sp	Spiral		E.F.	Each face
CT	Column tie		T	Top
E.W.	Each way		Bot.	Bottom

Figure 9·4 Symbols and abbreviations

B—beams
C—columns
L—lintels
S—slabs
J—joists
G—girders
W—walls
F—footings
T—ties
V—stirrups
D—dowels

SCHEDULES

The schedules for structural concrete components and their reinforcing appear on the drawings. These listings provide information as to the types of members and their size; the number, size, and location of reinforcing; and the number, size, and spacing of stirrups, etc. The schedules account for each piece of reinforcing, its specifications, and the bending (if required).

The actual location of schedules on the drawing sheet may vary, but in most cases the schedule is placed in the upper right-hand corner of the sheet. Notes are often located directly below the schedule. The schedules shown in Figure 9-5 are typical of a much-used type of schedule format.

DESIGN DRAWING PRACTICES

The preparation of design drawings should be done in such a manner as to provide all information necessary. Separate framing plans should be drawn for each floor and for the roof framing. Combinations of framing and architectural plans are often confusing. Even superimposing one framing plan on another can result in errors due to the increased chances for the drawings to be misread.

Each set of design drawings should contain the necessary framing plans, sections, and elevations. In addition, typical details showing the exact configuration of reinforcing should be prepared for a slab, joist, beam, column, girder, and other components included in the job. Particular care should be taken in clearly identifying size and arrangement of concrete structural members and the size, direction, bending, and termination points of the reinforcing. The drawings can provide a large amount of the necessary information, but they should be supplemented with schedules, notes, and references to recognized standards (where applicable).

Items of supportive information should be furnished in notation covering such data as live loads, soil bearing values, concrete quality, reinforcing steel stress values, fireproofing, etc. This type of data can prove extremely helpful throughout the balance of the detailing process and during the estimating, fabrication, and construction phases, as well. A portion of a typical design drawing appears in Figure 9-6.

PLACING DRAWING PRACTICES

The drawings for the fabrication and placing of reinforcing steel are not used in preparing the formwork for the concrete members. Their function is primarily that of clearly detailing all aspects of the reinforcing. The necessary dimensions on placing drawings are those used to detail the steel. The drawings, schedules, and

BEAM SCHEDULE

BEAM MARK	SIZE		REINFORCING STEEL			
			LONGITUDINAL		STIRRUPS	
	W	D	#Req'd. & Size	Remarks	#Req'd. & Size	Spacing from support face
3B1	14	20	2 - #5	Bot. - St.	10 - #3	3 @ 6 , 3 @ 11
			2 - #6	Bt.		
3B2	12	20	1 - #7	Bot. - St.	10 - #3	3 @ 6 , 2 @ 12
			2 - #6	Bt.		
3B3	8	12	2 - #5	Bot. - St.	12 - #3	3 @ 6 , 3 @ 10
			1 - #6	Bt.		
3B4	14	20	2 - #6	Bot. - St.	12 - #3	3 @ 7 , 3 @ 12
			2 - #7	Bt.		

COLUMN SCHEDULE

MARK	NO. REQ.	SIZE	REINFORCING STEEL			
			VERTICAL	TIES	REMARKS	
C1	12	18 x 18	4 - #7	#3 @ 24	See Section C-C	
C2	10	18 x 24	6 - #6	#3 @ 16	See Section C-C	
C3	6	24 Dia.	11 - #7	Sp. #4 @ $2\frac{5}{8}$	Splice dist. = 24 diameters	

Figure 9-5 Example of a beam and a column schedule

notes should clearly give the size, shape, bends, and locations of all bars, including how they are to be placed in the forms.

The reinforcement for beams, joists, and columns is shown in schedule form. The schedules should include number, mark, and size of the structural member; the number, size, and length of straight bars; the mark, number, size, total length, and bend details of bent bars and stirrups; spacing of stirrups and bar supports; and any other necessary information.

The reinforcement schedule for columns should include a supplemental schedule for column ties and bent bars, with diagrams showing the placement of these and the bending data for them.

Slabs are usually sufficiently detailed on the drawing so that the reinforcing is shown there rather than in a separate schedule. The slab system often contains a number of similar panels. These panels are assigned the same identifying marks and are detailed only once.

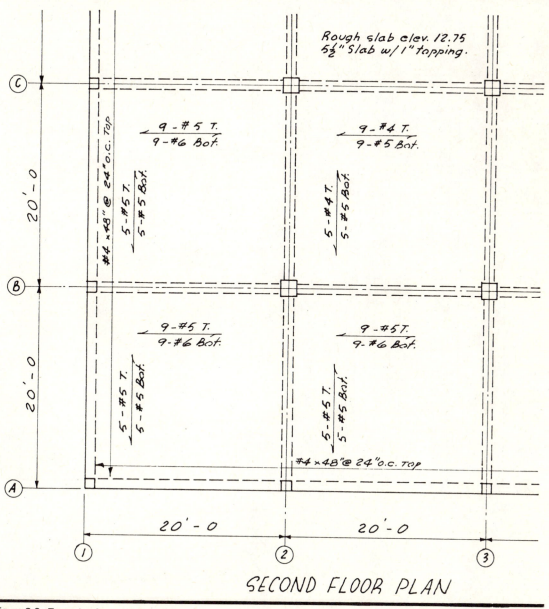

Figure 9·6   Example of a design drawing

Dowels are bars which are placed in one member and allowed to project so that a subsequent member can be poured later and joined to the first. These bars should be detailed with the initial member to preclude the possibility of their being left out. The partial drawing in Figure 9-7 represents a typical placing drawing.

BENDING OF BARS

The listing of bent bars in a concrete struc-

tural system can often be a complicated process. In an effort to simplify the process and facilitate understanding of the completed schedules, diagrams of the various bar configurations are assigned letters. Most fabricators have standard charts of typical bends and the designations assigned to each.

The most-used (and therefore standard) slant (or angle of slope) for bends is 45°. Bars may be specified for other slants, but it is much more advantageous to rely upon

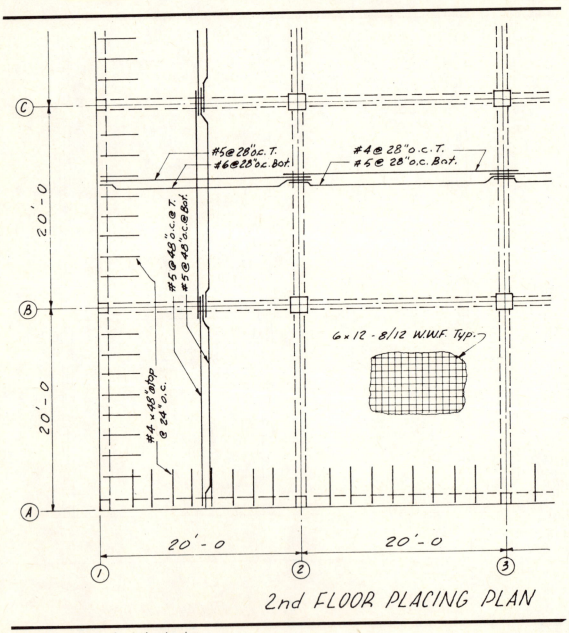

2nd FLOOR PLACING PLAN

Figure 9-7 Example of a placing drawing

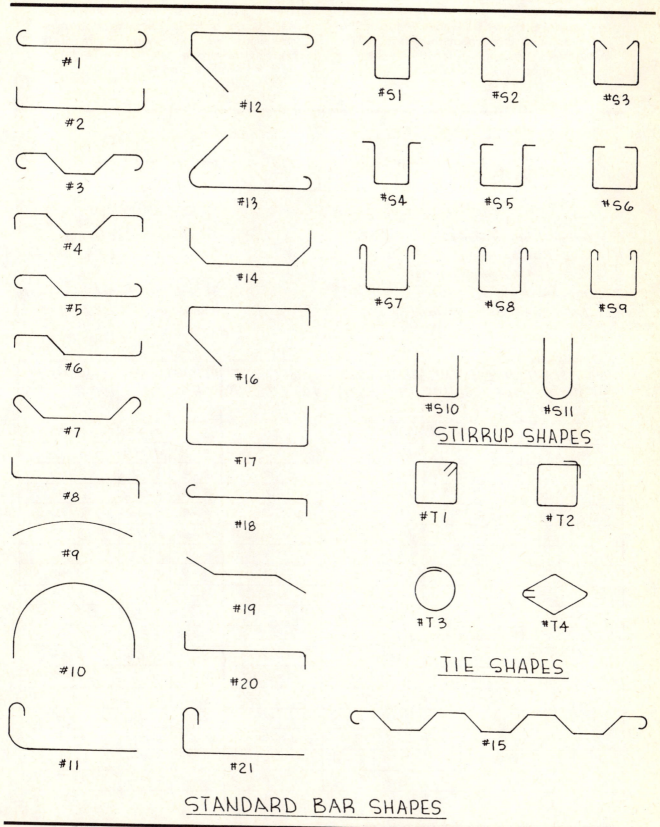

Figure 9·8 Typical bent bar shapes

the standard 45° slant wherever possible. Figure 9-8 shows a number of typical bar bends used throughout the industry. The diagrams shown are for the most part based upon the 45° bend.

The fabricator relies heavily upon another chart when bending bars, to determine dimensions required and to facilitate choosing the proper length of stock. This chart, shown in Figure 9-9, provides data

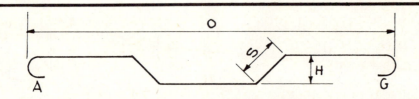

O = Overall bar dimension
H = Height of bar bend
S = Slant = 1.414 H to nearest $\frac{1}{2}$"
I = Increment = S - H
2I = Increment for two slants = 2 × (S-H)

Note: All dimensions are taken from out-to-out of the bar.

Scheduled total bar length should be = O + 2I + A + G.

HEIGHT "H"	SLANT "S"	INCREMENT FOR 2 SLANTS "2I"	HEIGHT "H"	SLANT "S"	INCREMENT FOR 2 SLANTS "2I"	HEIGHT "H"	SLANT "S"	INCREMENT FOR 2 SLANTS "2I"
			1-1	1-6$\frac{1}{2}$	11	3-1	4-4$\frac{1}{2}$	2-7
			1-2	1-8	1-0	3-2	4-5$\frac{1}{2}$	2-7
			1-3	1-9	1-0	3-3	4-7	2-8
2	3	2	1-4	1-10$\frac{1}{2}$	1-1	3-4	4-8$\frac{1}{2}$	2-9
2$\frac{1}{2}$	3$\frac{1}{2}$	2	1-5	2-0	1-2	3-5	4-10	2-10
3	4	2	1-6	2-1$\frac{1}{2}$	1-3	3-6	4-11$\frac{1}{2}$	2-11
3$\frac{1}{2}$	5	3	1-7	2-3	1-4	3-7	5-1	3-0
4	5$\frac{1}{2}$	3	1-8	2-4	1-4	3-8	5-2	3-0
4$\frac{1}{2}$	6$\frac{1}{2}$	4	1-9	2-5$\frac{1}{2}$	1-5	3-9	5-3$\frac{1}{2}$	3-1
5	7	4	1-10	2-7	1-6	3-10	5-5	3-2
5$\frac{1}{2}$	7$\frac{1}{2}$	4	1-11	2-8$\frac{1}{2}$	1-7	3-11	5-6$\frac{1}{2}$	3-3
6	8$\frac{1}{2}$	5	2-0	2-10	1-8	4-0	5-8	3-4
6$\frac{1}{2}$	9	5	2-1	2-11$\frac{1}{2}$	1-9	4-1	5-9$\frac{1}{2}$	3-5
7	10	6	2-2	3-1	1-10	4-2	5-10$\frac{1}{2}$	3-5
7$\frac{1}{2}$	10$\frac{1}{2}$	6	2-3	3-2	1-10	4-3	5-11$\frac{1}{2}$	3-6
8	11$\frac{1}{2}$	7	2-4	3-3$\frac{1}{2}$	1-11	4-4	6-1$\frac{1}{2}$	3-7

Figure 9-9 Typical bar bending data

on the height, slant, and the increment for two slants. The information is most useful in laying out bends and in figuring proper length of material prior to bending. Such charts are available covering a number of the various bar configurations.

The American Concrete Institute (ACI) has published standards for the various aspects of concrete construction. The bending charts are only one example of the type of material provided in the standards manual. The reinforcement of columns

1 - TIE
4 - BARS

2 - TIES
6 - BARS

2 - TIES
8 - BARS

3 - TIES
10 - BARS

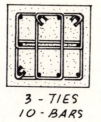

3 - TIES
12 - BARS

BAR SIZE (Verticals)	SIZE & SPACING (in inches) OF TIES. Max. spacing to be less than least col. dim.		
	#2 bar	#3 bar	#4 bar
#5	10	10	10
#6	12	12	12
#7	12	14	14
#8	Not recomm.	16	16
#9	" "	18	18
#10	" "	18	20

Figure 9·10 Typical column ties and spacing chart

COLUMN DIAMETER (inches)	SIZE OF SPIRAL	MAX. # & BAR SIZE					
		#5	#6	#7	#8	#9	#
14	3/8	12	11	10	9	7	
15	3/8	13	12	11	10	8	
16	3/8	15	13	12	11	9	
17	3/8	16	15	14	12	11	
18	3/8	18	16	15	13	11	
19	3/8	19	18	16	15	13	
20	3/8	21	19	18	16	14	
21	1/2	22	20	19	17	15	
22	1/2	23	22	20	18	16	

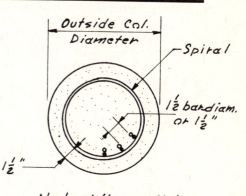

Outside Col. Diameter

Spiral

1½ bar diam. or 1½"

1½"

Note: When splicing vert. bars, bars from below are inside of bars above & use a common radius.

Figure 9·11 Column bars for round columns

requires both vertical steel and bars tying the vertical steel together. Such ties may be separate horizontal units or a continuous spiral of • steel bars. The material shown in Figures 9-10 and 9-11 are ACI recommendations for the tying of column reinforcing.

Column ties must be bent from straight bar stock. The previous bending chart of Figure 9-8 provides diagrams of the various shapes necessary for column ties. Notice that four basic types were shown, labeled T-1 through T-4.

SUMMARY

Dating back some 2,000 years, concrete has been the basis of countless structural systems. Through experience in its use, and technological advances, the material provides excellent structural capabilities for today's structures. Based upon natural elements and materials, concrete possesses long-lasting structural qualities whether in poured-in-place or precast form. The prestressing of concrete members now permits reduction in the overall member size and weight.

The drawings necessary for detailing concrete structural systems fall into two main categories. The design drawings provide the information necessary to lay out, form, and construct the system. Placing drawings furnish data necessary to fabricate and place the very necessary reinforcing steel.

With excellent quality control, and modern engineering practices, concrete continues to grow in use, and in the sophistication of its use.

Problems – Chapter 9

1. Prepare a list of the various types of cement, aggregates, and reinforcing available from building material firms in your area.
2. Prepare simple sketches for a garden footbridge in two construction methods. Base one design on a cast-in-place system and the other on a precast system. Specifications for the concrete bridge and railings are:
   Span = 6 ft 0 in.
   Width = 3 ft 0 in.
   Height of side railings = 2 ft 6 in.
   Thickness of slab = 3 in.
   Indicate in the sketches the procedure you would follow to form and erect the bridge.
3. Prepare pictorial drawings (small) of the various standard types and shapes of precast concrete blocks.
4. Create a table of standard deformed reinforcing bars, giving their size and proper designation and symbol.
5. Gather manufacturers' literature describing the various precast and prestressed structural units currently available in your location.
6. Create a single sheet showing the standard lines used in drawing for structural concrete and the various symbols and abbreviations (with their meanings) used on such drawings.

**After Studying This Chapter
You Should Be Able To:**

1. Describe the most frequent use of concrete in residential construction.
2. Cite the types of experiments which have been tried in attempts to create total concrete dwellings.
3. Name the functions of footings and foundation walls.
4. Explain why soil bearing values must be considered in the design of footings.
5. Describe how concrete *shears* and how the shear pattern affects the design of footings and their reinforcement.
6. Describe the possible effect of not having foundation walls centered on footings.
7. Define *pad footings*, and explain their use.

16. List factors which must be included in an adequate wall design for a concrete-block residence.
17. Describe various methods of constructing residential floor slabs, including steps taken to guard against moisture and cracking.

INTRODUCTION

Residential construction has relied upon concrete in some manner for many years. The most widespread use of concrete in residences was for footings, foundation walls, and on-grade or below-grade slabs. Such use continues to be the most prevalent application of concrete in the present housing industry.

The use of concrete masonry units (blocks)

---

chapter 10
concrete structural systems-residential

---

8. Describe the factors which the designer considers in determining how deep to set the footings.
9. Describe the pressures which basement walls must withstand.
10. Describe the typical reinforcing system used in foundation walls in residential construction.
11. Name the steps which can be taken to keep foundation walls from shifting on the footings.
12. List important practices used in the construction of concrete-block foundation walls and explain how such walls are reinforced.
13. Explain the differences between most residential drawings and the design and placement drawings prepared for commercial and industrial structures.
14. Describe the four main methods of edge construction for slabs in residential slab-on-grade construction.
15. Define a *grade beam*, and list its uses.

has been popular in constructing residences in some regions of the country. These blocks provide structural and dividing partitions and walls without additional structural components.

Experiments involving total structural systems of concrete for houses have taken place for a number of years. Some experiments used formwork and actually poured concrete in place for floors, walls, and roof slabs. Other experiments were pointed toward precasting units on site and raising them to form walls, upper floors, and roofs. Still other schemes relied upon poured-in-place or precast columns and beams, with the actual walls, floors, and roofs made up of other materials.

New structural applications in residences have been based upon the development of prestressed and lightweight concrete components. The standard members and units available are of reduced size, have better surface finishes, and lend themselves to the residential scale much more effectively than their predecessors.

FOOTINGS

The footings of a structure provide the necessary support for the entire construction. Their purpose is to spread the loads from the structure over enough area to permit the soil to support the building. Various soils have their own bearing capacity, and the designer determines the values from test borings and material tests. If footing areas are not of sufficient size, the building may settle, causing cracking of foundations, slabs, and walls.

Footing sizes Most footings are wider than they are tall and are of rectangular cross section. Since concrete shears at a 45° angle, the footing is most often a minimum of twice the width of the foundation wall it supports. Footings should be designed for the specific load and soil conditions, but as an example for residential work, 8-in. foundation walls require 16-in.-wide footings, 10-in. walls take 20-in. footings, etc. The depth of footings varies, but

because of the 45° shear they are usually one-half as deep as they are wide.

Footing designs The illustration in Figure 10-1 shows some typical footing designs, with the dotted line showing the portion which would be of little value if actual shear conditions prevailed.

One type of footing, shown in the illustration, takes for granted that the areas outside the shear line are of little value (upper corners) and eliminates them. These footings, with sloping sides (45°), are often called *bell* footings because of their cross-sectional shape.

Footing reinforcement The placement and specifications of reinforcing in residential footings varies with the particular design requirements. It is common, however, to avoid placing reinforcing bars in the area (upper corners) which might be separated by shear stresses, since after shear failure the rods would no longer serve their intended purpose.

Residential footings (and most other footings) are designed in such a manner that they will be centered on the foundation walls which they support. This practice as-

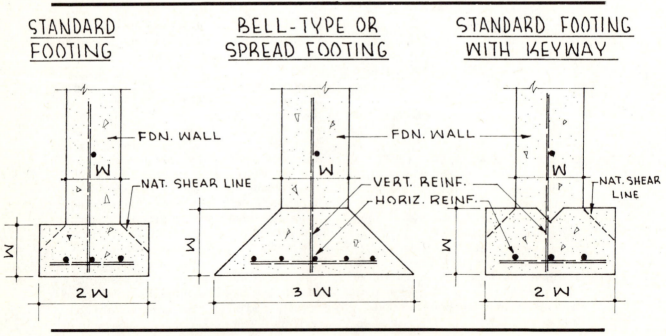

Figure 10·1 Typical residential footings

sures that loads will be transmitted directly down onto the footing. When footings are not centered with foundation walls, eccentric loadings can occur which have a tendency to tip the footing. The same condition can create greater pressures on the bearing soil under one side of the footing.

Pad footings Basement columns, porch posts, and other vertical components require footings to distribute the load they transmit. Footings for this purpose are often of the pad variety, consisting of a square or rectangular slab. The sketches in Figure 10-2 show typical footings for such applications.

Reinforcing in pad footings usually consists of bars running in two directions. The bars actually form a reinforcing grid and would be called out in the manner of *3-#5 each way.*

DEPTH OF FOOTINGS

It is necessary to excavate deeply enough to reach the level of undisturbed soil with sufficient bearing capacity. The footings for basements are generally at such a level as to permit pouring them directly below basement floor slab depth. Shallow footing depths for porches, crawl-space construction, or slab-on-grade construction should also reach undisturbed soil.

When footings lie in the soil depth subject to freezing and thawing, great damage can result. Moisture in the soil causes it to expand and contract under such conditions, thereby creating sizable pressures on footings and walls. For this reason, footings are placed at depths below the freezeline. The depth to which freezing of the soil occurs varies with climatic areas of the country. The designer must be aware of the freezeline for the particular location, if he is to guard against the failure of footings due to the resulting expansion and contraction.

FOUNDATION WALLS

The foundation walls transmit the loads of the structure to the footings, which in turn transmit and distribute them to the soil. Foundation walls also encounter the horizontal loads exerted by the soil adjacent to them. In the case of basement walls, the magnitude of the horizontal pressure from the soil can be quite large. Design of the

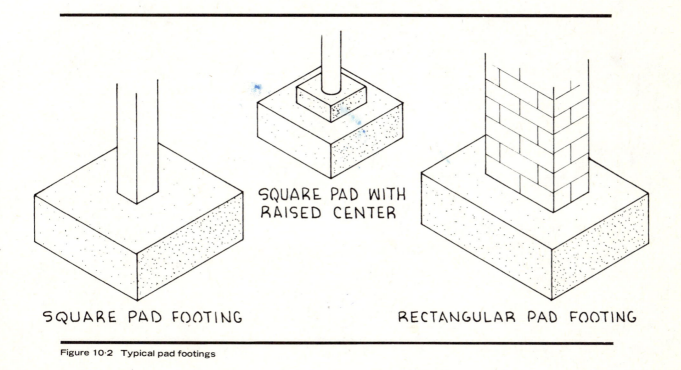

SQUARE PAD WITH RAISED CENTER

SQUARE PAD FOOTING

RECTANGULAR PAD FOOTING

Figure 10·2 Typical pad footings

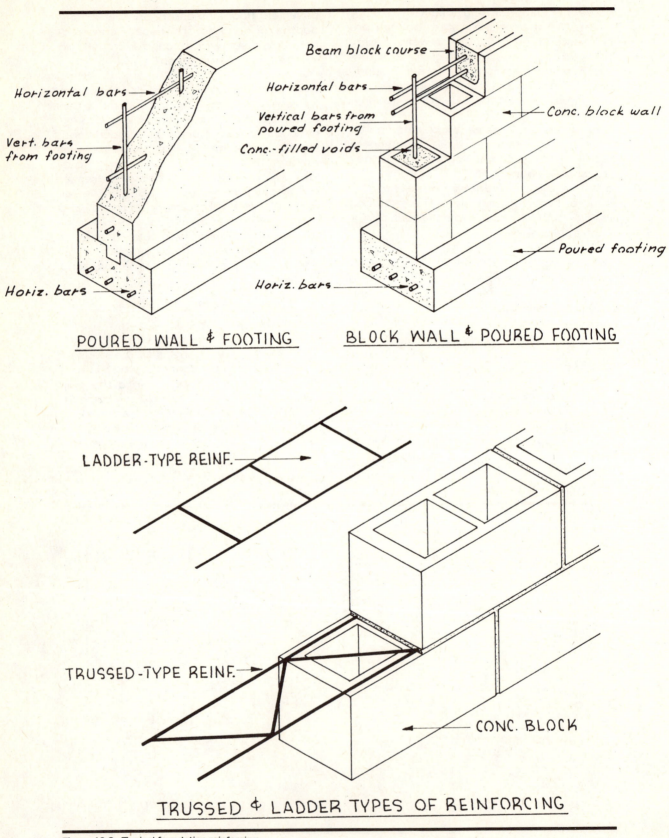

POURED WALL & FOOTING

Horizontal bars

Vert. bars from footing

Horiz. bars

BLOCK WALL & POURED FOOTING

Beam block course

Horizontal bars

Vertical bars from poured footing

Conc.-filled voids

Conc. block wall

Poured footing

Horiz. bars

LADDER-TYPE REINF.

TRUSSED-TYPE REINF.

CONC. BLOCK

TRUSSED & LADDER TYPES OF REINFORCING

Figure 10·3 Typical foundation reinforcing

walls must be adequate to withstand both the vertical loads from above and the horizontal loads from outside.

Residential foundations are specified with various thicknesses, but, most frequently, basement walls range from 8 through 12 in., and foundations for porches, planters, etc., may be as thin as 4 in.

Foundation reinforcement Reinforcing of foundation walls generally consists of bars running in both horizontal and vertical directions. The horizontal bars are spaced rather closely together (18 in., 24 in., etc.), with vertical bars spaced at 36 in., 48 in., etc. The horizontal bars encounter some of the stresses, while the vertical bars provide continuity for them by tying them together and provide stiffness to resist horizontal pressures on the wall and footing. Since the vertical bars are placed in the footings as they are being poured, they provide a means of tying the two pours (footings and foundation) together.

Keyed footings and foundations Keyed footings provide another means of ensuring that the foundation will not separate from, or move on, the footing. A slot or keyway is created in the top of the footing, and concrete from the foundation wall enters the keyway, forming a locking device which precludes any horizontal separation between the two.

Concrete-block foundations Foundation walls in residential construction are often constructed with concrete-block units laid in much the same manner as brick. The principles used in reinforcing concrete-block walls are essentially the same as those used for poured-concrete walls. Both horizontal and vertical reinforcing is provided.

Reinforcing for block walls Special concrete blocks are produced which permit running horizontal bars through them and filling the continuous void with concrete. These blocks are referred to as *beam* blocks. Foundation walls do not require beam blocks at every course, so the main wall is built with regular blocks, and the beam blocks with bars occur at spaced intervals, including the top course.

Another form of horizontal reinforcing is shown in Figure 10-3. This item consists of steel wires running continuously at the sides, connected with diagonal wires (welded). Placed continuously at intervals between courses, the mesh provides horizontal reinforcing.

Vertical reinforcing of block walls is accomplished with vertical bars (with ends placed in the footing) running through the voids in the blocks. The voids are poured full of concrete, resulting in a continuous vertical structural element.

DRAWINGS FOR RESIDENCES

Since the predominant use of concrete in residential construction is for footings, foundations, slabs, and concrete-block walls, this section deals with those applications. The drawing practices for other concrete structural systems for residences conform to those used in commercial work and are therefore covered in the next chapter.

Residential drawings are often less detailed than commercial or industrial plans, and several purposes may be combined on a single drawing. The structural drawings for a residence, in this case footings and foundations, include a footing and foundation plan which may very well be a basement plan as well. In addition to such a plan, the set of drawings includes all sections and details necessary to describe adequately what is to be done.

FOUNDATION PLANS AND DETAILS

The typical foundation plan is a drawing of the entire foundation system. The foundation walls are drawn with heavy object lines. The footings are shown on the same plan with broken (hidden) lines. Such features as recesses, pockets, etc., are drawn with solid or broken lines, depending upon their location.

Foundation plans are actually horizontal sections cut through the foundation wall system. If there are foundation vents, or basement windows, the cutting plane is most often taken at a height allowing these features to be drawn with object lines. This is similar to the procedure used in drawing floor plans.

Even though footings are drawn with hidden lines, the relative heights can be indicated. The lines of the deeper footing give way to the lines of a footing above it. The situation of a 30-in.-deep porch footing crossing the basement footing at 90° to intersect the basement wall is a good example. Figure 10-4 shows that it is clearly the porch footing which is above the basement footing.

Foundation plans for residences generally include notes and items pertaining to the floor system above. The foundation plan dimensioning is done in such a manner as to indicate overall dimensions, distances to other features such as recesses or windows, the width of foundation walls and footings, and the location of columns, piers, pad footings, etc.

Examples of foundation plans The foundation plan in Figure 10-5 is for a frame residence with a crawl space below the floor. It contains the necessary information about the footings, foundations, piers, and the intended structural system for the floor. Notice that footing and wall reinforcing is shown and called out on the sectional details, rather than on the plan or in a schedule. This is a common practice in residential work.

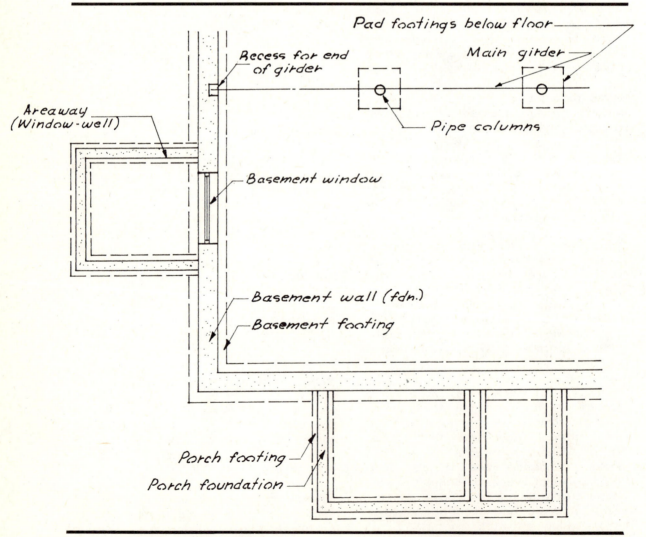

Figure 10·4 Partial footing and foundation plan

Similar to the plan shown in Figure 10-5 is the basement plan of Figure 10-6. This drawing is prepared for a frame and brick veneered residence with a full basement.

The drawing procedure is the same as the previous example, and this plan includes information on the stairs, floor joists, beams, windows, and areaways (window wells).

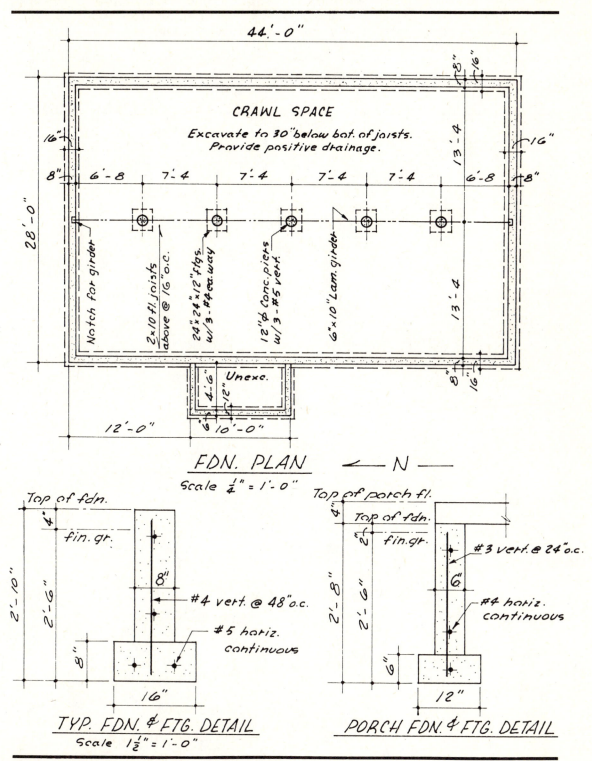

FDN. PLAN   ← N —

Scale ¼" = 1'-0"

TYP. FDN. & FTG. DETAIL

Scale 1½" = 1'-0"

PORCH FDN. & FTG. DETAIL

Figure 10·5 Foundation plan for crawl space of house

Concrete-block walls would be drawn in a similar fashion with only the material symbol (cross-hatching) being different.

Details for slab construction Many residences are constructed with concrete floor slabs on the grade (earth-supported). There are some four main methods of construction used in such instances. Figure 10-7 provides sectional details of each of

the four methods. It is easily seen that foundation configurations vary with the method used. Detail (A) shows a system which isolates the floor slab from the foundation system. Detail (B) has a recess to allow the floor slab to bear on one-half of the foundation wall.

The system in detail (C) permits the floor slab to run completely over the foundation

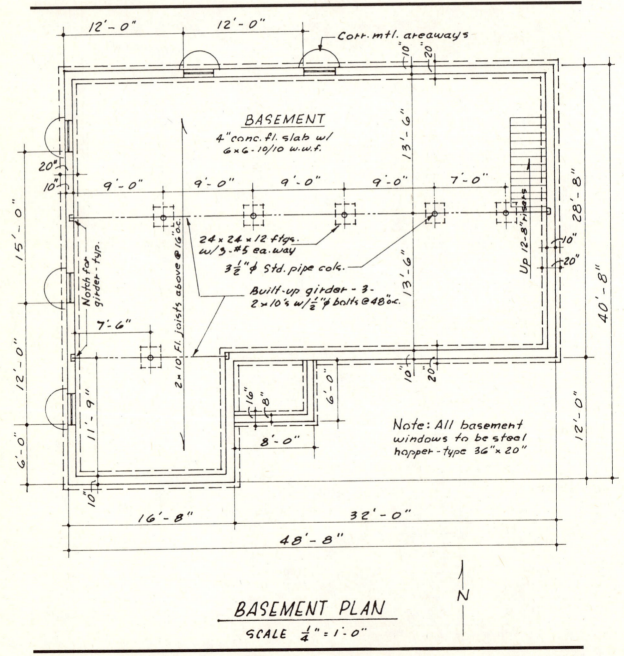

Figure 10·6 Footing and foundation (basement) plan

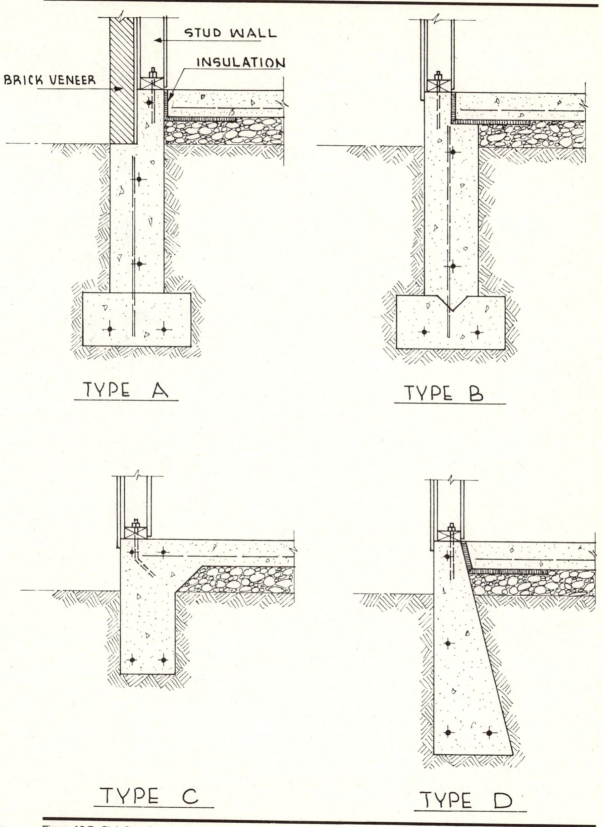

STUD WALL

INSULATION

BRICK VENEER

TYPE A

TYPE B

TYPE C

TYPE D

Figure 10·7 Slab floor foundation types

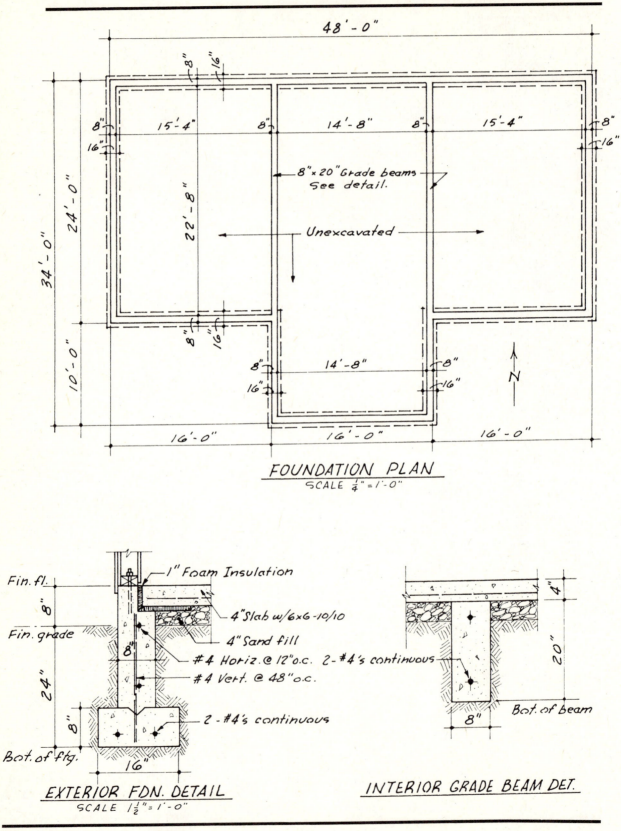

Figure 10·8 Foundation plan—slab floor

beam and be combined with the foundation system in one pour. The method shown in detail (D) involves an angular surface on both foundation and slab.

The use of a foundation without a footing, as in details (C) and (D), is not uncommon. In such a configuration, the foundation wall serves as a *grade beam*. Grade beams are also used for interior support of floor slabs and interior walls.

Example of slab on grade A typical foundation plan for a slab-on-grade frame residence appears in Figure 10-8. This particular residence relies upon the system shown

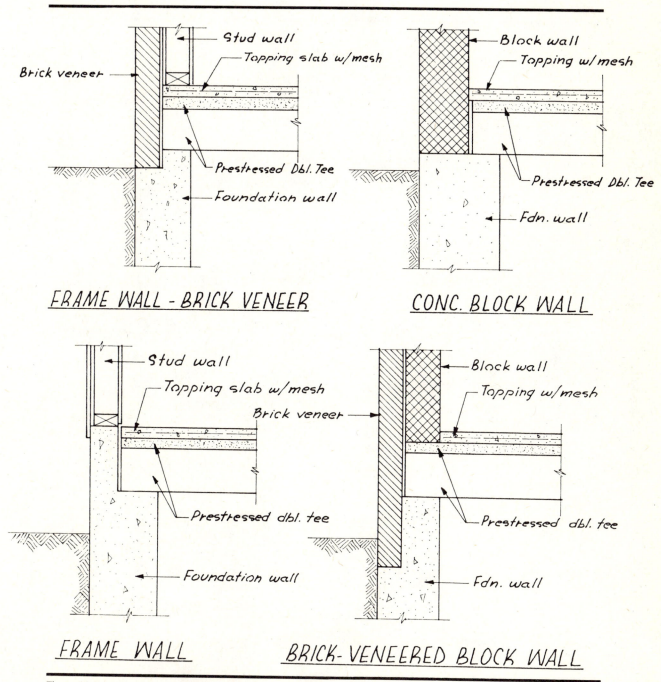

Figure 10·9 Typical end conditions – double T's

in detail (B) of Figure 10-7. In many areas of the country it is most important to provide insulation at the perimeter of the floor slab. The plan allows a 4-in. ledge to support the slab, which provides for 3 in. of bearing and a 1-in. space for rigid foam-board insulation.

When floor slabs, whether basement,

garage, or ground floor, are contained within a foundation (poured against), it is good practice to provide some type of expansion joint. Insulation can serve this purpose, as can the various types of manufactured expansion strips. The expansion strips provide a flexible cushion and help to prevent cracks in slabs and walls from expansion and contraction.

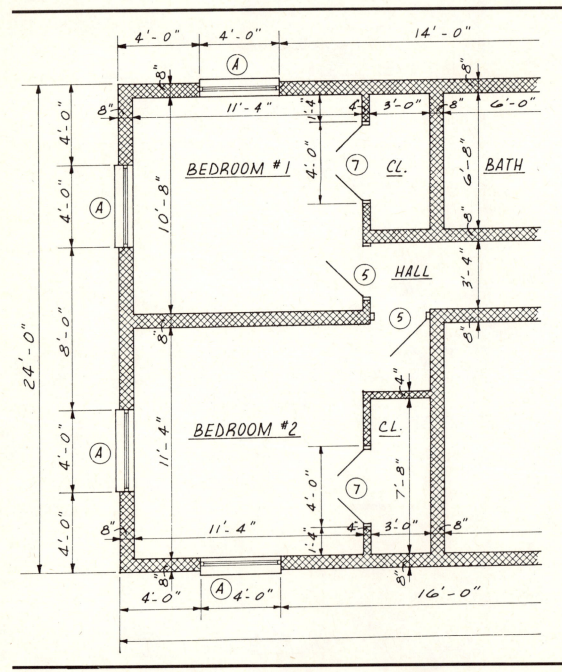

Figure 10·10 Partial plan—concrete block

Details for support of prestressed floor slabs The foundation systems shown in the previous examples are often found in a somewhat similar form in some commercial and industrial applications. The use of prestressed floor units is applicable to residential, commercial, and industrial structures. The rough sections in Figure 10-9 show various methods of supporting prestressed double T's for residential floor use. Foundation plans for such applications would be very similar to those explained previously.

CONCRETE-BLOCK CONSTRUCTION

Concrete-block walls for residential construction can be either load-bearing or non-load-bearing. Different types of mixture are often used in producing blocks for the two types of use. Lightweight blocks are usually used for dividing partitions which have no load-bearing requirements.

The floor plan for concrete-block residences is similar to that for any other type of construction. The block walls are shown on the architectural floor plan most often, rather than on a separate structural plan. It is important that the dimensions of the residence permit using block without cutting wherever possible. Modular planning, based upon the size of the block, greatly assists in assuring as little cutting as possible. The partial floor plan of Figure 10-10 shows a typical floor plan for a block residence.

Lintels and beam blocks It is imperative in block construction that lintels (above openings) and continuous bands of reinforced beam blocks be specified. The sectional details in Figure 10-11 clearly indicate the desired use and location of these special units. Notice that reinforcing mesh is called for at two-course intervals and that a continuous reinforced beam-block course is to be used at the top of the wall.

Modular heights In designing block wall construction, strict adherence to the modular heights of the blocks is an economic necessity. Both total wall heights and the height of wall openings should be based on the standard increments of block units. Nonconformance to standard increments can result in a huge amount of labor in cutting blocks for an entire course encompassing the total wall distance.

Reinforcing block walls The details in Figure 10-11 include notes and indications calling for vertical bars at 48-in. intervals. These bars would run through the hollow core (or voids) in the blocks, and concrete would be poured into the voids around the bars. Several vertical bars placed in this fashion may be required at exterior corners and intersections of walls.

CONCRETE FLOOR SLABS

Residences of all types of construction often require concrete slabs for the floors of garages, carports, porches, or basements. The edge details of such slabs can be varied and may conform closely to the details shown in Figure 10-7. Earth-supported floor slabs should have an adequate bearing surface beneath them. To assure a stable base, most slabs are placed on a sand or gravel fill. The thickness of the fill varies but often falls in the 2–4 in. range.

Vapor barriers The earth below the slab is often graded to provide as level a surface as possible. The sand or gravel is placed and is compacted and leveled. The slab may be poured directly on the sand, but many installations use a plastic (polyethylene) sheet between sand and slab. The plastic sheet serves as a vapor barrier and prevents moisture from entering the slab from below.

Slab reinforcing Floor slabs vary in thickness, but a 4-in. slab is often used. The concrete slab should be provided with the welded wire fabric (or mesh) mentioned in the previous chapter. The mesh can prevent cracking of the slab, particularly through expansion and contraction. Slabs which are not supported by adequate bearing may require reinforcing bars in addition to (or in lieu of) the mesh.

Scribing Exterior slabs are often more

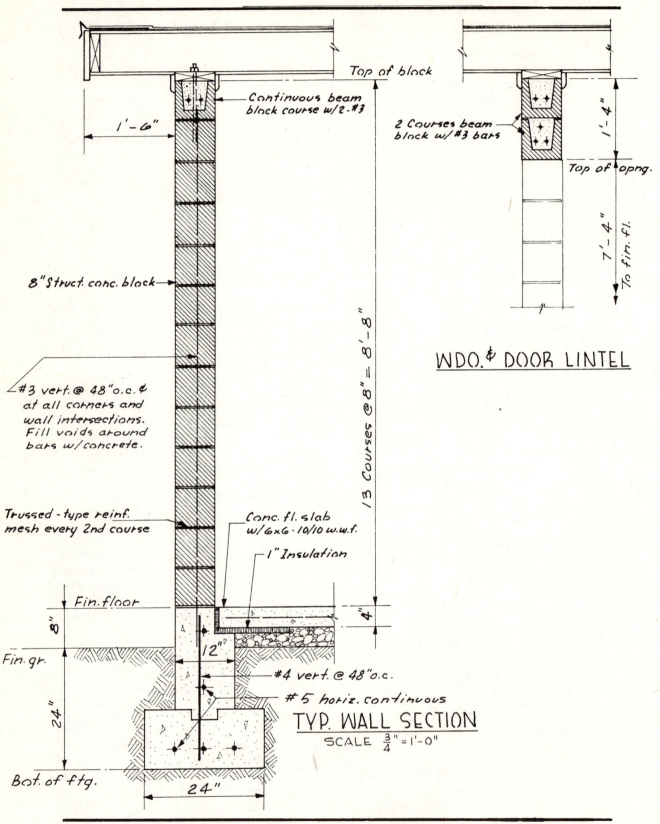

Top of block

Continuous beam
block course w/2-#3

2 Courses beam
block w/#3 bars

Top of opng.

1'-4"

7'-4"

To fin. fl.

WDO. & DOOR LINTEL

8" Struct. conc. block

13 Courses @8" = 8'-8"

#3 vert. @ 48" o.c. &
at all corners and
wall intersections.
Fill voids around
bars w/concrete.

Trussed-type reinf.
mesh every 2nd course

Conc. fl. slab
w/6x6-10/10 w.w.f.

1" Insulation

Fin. floor

4"

8"

Fin. gr.

12"

24"

#4 vert. @ 48" o.c.

#5 horiz. continuous

TYP. WALL SECTION
SCALE 3/4" = 1'-0"

Bot. of ftg.

24"

1'-6"

Figure 10·11 Typical block details

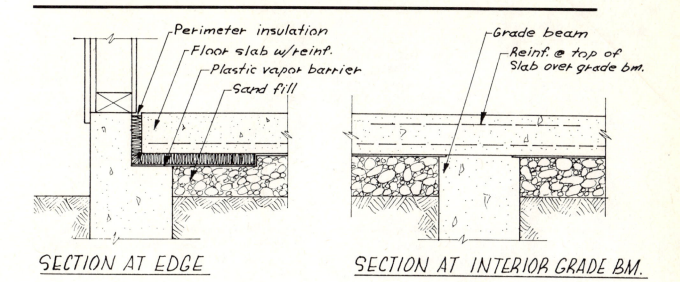

SECTION AT EDGE

SECTION AT INTERIOR GRADE BM.

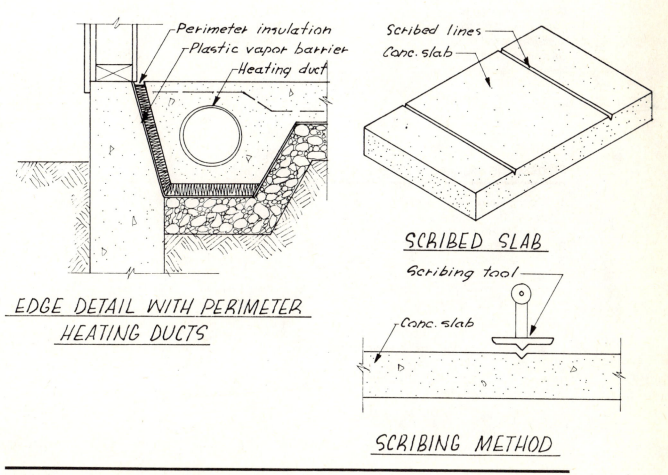

EDGE DETAIL WITH PERIMETER
HEATING DUCTS

SCRIBED SLAB

SCRIBING METHOD

Figure 10·12 Typical floor slab and slab scribing details

prone to crack than those which are enclosed. Since cracks are unsightly in any slab exposed to view, measures can be taken to keep the cracks where they will be least noticeable. Scribing of slabs is a much-used practice. This process entails creating grooves in the slab before it has hardened. The grooves may run in parallel lines, or patterns may be formed, such as rectangles, squares, etc. Cracks in the slab will be most apt to follow the grooves, thereby remaining unseen. The scribed patterns can serve a decorative, as well as a functional, purpose.

Figure 10-12 contains additional details of floor-slab construction and illustrations of the scribing practice just mentioned.

SUMMARY

The use of concrete in residential construction continues to be most applied to footings, foundations, and floor slabs. The largest use of concrete for residential wall construction is that of concrete blocks.

Residential applications of structural concrete usually require much less sophisticated drawings than do commercial and industrial structures. Structural concrete features are often called out on other drawings and total wall sections. Rather than complete schedules for reinforcing,

the desired reinforcing is most often noted on a wall section or floor plan. The combination of footing, foundation, and basement elements in a single drawing is not at all unusual.

Extensive concrete structural systems for residences utilize the same procedures used for commercial and industrial buildings, and those detailing procedures are therefore contained in the following chapter.

Problems—Chapter 10

1. Design and produce the drawings necessary for the foundation, footing, and slab for a 24 × 28 ft frame garage. (Assume that the frostline is 30 in. and that the top of the floor slab is 4 in. above the finish ground level.)
2. Determine the frostline for your geographic area.
3. Find a floor plan for a frame residence in a plan book or magazine, and prepare the foundation plan and necessary details for the footings and foundations.
4. Prepare the plans and details necessary for the concrete portions of a concrete-block vacation cabin. The size overall is to be 24 × 40 ft, with room arrangements of your choice.
5. Redesign (in sketch form) the cabin in problem 4, changing it to a frame wall system, and include a basement. Prepare the basement plan and necessary details for the footings and foundation.

After Studying This Chapter You Should Be Able To:

1. Name three main types of structural framing systems which are employed in commercial building construction.
2. Explain the importance of commercial buildings which permit ease of interior remodeling.
3. Cite advantages of prestressed concrete units for commercial structures.
4. Describe the attributes of concrete systems in regard to fire damage and maintenance.
5. Describe steps which can be taken to provide more easily for openings in concrete decks and for attaching ductwork, piping, etc.
6. Explain the importance of the detailer's projecting his thinking to the actual on-site construction process.

more than any other area of the construction industry during the last decade, followed closely by the industrial structures. Commercial structures, by their very nature and function, usually permit the designer a far broader opportunity for creative design. Some of the most exciting and innovative uses of structural concrete systems have occurred in commercial structures.

Commercial structures utilize concrete for each of the three main structural framing types. Wall-bearing, beam-and-column, and long-span framing systems employing concrete components are found frequently in commercial buildings. The attribute of these systems in providing large areas of open space is most useful in commercial construction. Commercial buildings are often remodeled as space needs fluctuate,

chapter 11
concrete structural systems- commercial

7. Describe drawings necessary for concrete structural systems.
8. Name the factors which affect the design of footings and foundations.
9. Describe methods used in providing adequate foundations on steeply sloping sites and on projects where the bearing values of the soil vary greatly within the area of the building.
10. Describe several of the various types of footings and foundation systems.
11. Define *grade beams*, and explain the functions they serve. Define *dowels*, and explain their purpose.
12. Describe *marking* systems for concrete structural members, and cite their importance.
13. Describe *engineering* (design) drawings and *placing* drawings, and cite their similarities and differences.

particularly in removing and creating interior partitions. Systems which preclude the necessity of a maze of load-bearing interior walls are of great advantage in this respect.

The development of standardized prestressed and precast concrete structural members has fostered increasing use of these components for commercial construction. Rigid standards followed by the precasting industry have resulted in well-finished surfaces which, when left exposed to view, are appropriate for a commercial building.

Although many structures still use a poured-in-place system, precast and prestressed units permit a much more rapid erection process, which generally results in economic advantages. Not only can labor and time be saved directly, but often the building owner can realize great savings by occupying the new structure as quickly as possible.

INTRODUCTION

The field of commercial buildings has probably utilized concrete structural systems

Concrete structural systems, when properly designed and treated, carry excellent fire ratings. Commercial firms often carry extensive stocks of products or have extremely important files of records and documents. The need to guard against destruction of such items, and the safety of employees and clientele, make well-designed concrete structures an attractive solution to the building problem.

MAINTENANCE FACTORS

Maintenance of commercial buildings can become a very demanding process for the owners. The structures are often subjected to heavy traffic and its accompanying problem of upkeep. Many materials require constant maintenance, and dollars spent to achieve proper care mean less profit for the firm. Concrete structural systems provide relative ease of maintenance. Exterior precast wall panels of the exposed-aggregate type eliminate a great deal of painting or maintenance of the joints of brick and stone work. Rust does not affect concrete as it does steel, and the weathering problems associated with wood are practically eliminated with concrete systems.

APPEARANCE FACTORS

The actual design of concrete structures for commercial purposes must constantly be viewed with an eye toward appearance. Joints and connections may very well be exposed to view, and practices which might be adequate in industrial construction are often unsightly when placed in a commercial context. Many commercial buildings use a sprayed-on acoustical treatment applied directly to the underside of concrete decks, prestressed units, etc. Protruding joint hardware or improperly designed joints can ruin the intended effect of such applications.

OPENINGS FOR AND ATTACHMENT OF EQUIPMENT

Provisions must be made for ductwork, wiring, piping, and other mechanical systems and devices. It is often possible to provide attachment or hanging devices which are placed in the concrete members as they are being poured. Careful planning of structural units can facilitate better and faster installation of other necessary components. A number of devices are manufactured for placement in concrete units and are available in both projecting and recessed types.

It is possible to create openings in concrete units after their fabrication. Many installations require that openings for vertical ductwork, piping, and wiring be cut after the structural units are in place. Advanced planning can provide for many such openings to be formed as the units are being cast, and the forming process is far more efficient than the cutting required after the units have been created. This problem is of much greater magnitude when dealing with the heavy structural units. The lightweight roof slabs can be much more easily altered on the job site.

STRUCTURAL CONNECTIONS

The combination of structural concrete systems and deck and wall panels, or of other wall, floor, and roof materials, requires some provision for attachment to the structural system. It is true that many such attachments can be made after the fact, but it is far more logical and economical to attempt to foresee all fastening problems in the design and detailing stages. Chapter 9 described several methods of providing for the attachment of panels, etc. Most used, of course, are systems involving bolting or welding, with the actual fastening devices embedded in the individual concrete units as they are being poured.

DESIGNING AND DETAILING

As with any other building system or type, the design and details for a concrete commercial building must be accurate, logically planned, adequate, and complete. Completeness, however, does not negate the need to be concise as well.

The structural detailer must make sure that his work conforms to standards and practices which are accepted throughout the industry and which impart the same meaning to all of those working on the various facets of the project. He must project his thinking to the various stages of construction in the field. It may be easy to create a detail which looks good on paper, but the detailer must realize that it must also work well (and be feasible) during the actual construction process on the job.

DRAWINGS

The necessary drawings and drawing procedures for concrete structural systems were explained in the first chapter of this unit. Commercial buildings should be detailed in the same manner. The poured-in-place structures should have both engineering and placing drawings with their accompanying details and schedules.

Very simple structures do not require the same sets of drawings as do those with more elaborate structural systems. Actual placing drawings (and some engineering drawings) can often be eliminated when components are standard precast or prestressed units. In such cases, the drawings necessary to produce the units are prepared by the fabricator to the requirements adopted as standards by his particular industrial association. The preparation of drawings for production by the manufacturer does not eliminate the need for the framing and erection details needed to construct the building at the site.

The balance of this chapter focuses on the various structural types and the detailing necessary for their construction. Since all of the various system types depend upon a footing and foundation system, the information regarding footings and foundations is presented in the next few pages, rather than appearing in part with each system type.

FOOTINGS AND FOUNDATIONS

The chapter on residential systems spoke of the necessity of providing adequate footings and foundations to support the structure. Certainly the support is necessary for all types of buildings. The adequacy of such supportive elements becomes even more important when dealing with commercial and industrial structures. These buildings are usually larger structures to begin with and rely upon long-span components which greatly increase the pressures exerted upon foundations directly beneath their end-bearing points. The increased heights of wall materials add measurably to the loads transmitted to footings and foundations.

Since loads on the foundation and footings of commercial buildings are much larger and more concentrated, greater efforts are made to reach adequate bearing soils. By like token, more care is taken to ensure that loads are dispersed sufficiently to permit the soil to support them. It often becomes necessary to excavate to different depths for footings in a single project, in order to obtain satisfactory bearing for all portions of the structure.

Footings at varying depths Sloping sites and split-level types of structures create additional areas of concern in the design and detailing of footings and foundations. In an effort to reduce material and labor costs, buildings in such cases may utilize *stepped* footings. These footings descend the slope at intervals, with the foundation walls varying in depth as required to bear on them. A single wall could easily have its footings at four or five different depths.

The alternative for stepped footings on a sloping site is to excavate all footings to the same depth as the deepest footing on the job. Such a practice results in much more wall construction, and the extreme depth of footings on the high side can create difficulties in excavation, forming, and placing operations.

Varying soil conditions Soil conditions can vary within the area of a building, and the only sure means to determine accurate bearing values is to take test borings. These tests are taken at a number of points within the area to be covered by the structure (usually around the perimeter of the proposed building).

When soil conditions suddenly change, and small areas of the footings will be affected, it is not uncommon to support the footings in turn with concrete piles or piers. Holes are drilled from regular footing depth to deeper depths to provide bearing on rock or better soil below. The piles (or piers) are poured before (or with) the footings and contain whatever reinforcing is necessary. These extra footing elements occur at intervals such as 6 ft, 8 ft, 10 ft, or the distance for which their bearing was designed.

Eccentric loads Mention has been made previously in the text of the fact that footings and foundations may be called upon to resist stresses of an eccentric or side-thrust type. Walls of basement or below-grade spaces encounter side pressures from the earth outside. Off-center loads applied to footings tend to tip them. Some structural systems create their own high stresses. Such components as arches can generate and transmit extremely high stresses in an outward direction at their base. Footings and foundations for these units must often be of the *buttress* type to resist the stresses properly, or be tied together with reinforced grade beams (in essence tying the arch legs together).

TYPICAL FOOTING AND FOUNDATION DRAWINGS (SIMPLE STRUCTURES)

The same practices explained in Chapter 9 apply to preparation of drawings for footing and foundation systems for commercial structures. The actual foundation walls are most often indicated with solid object lines, while footings are shown with broken hidden lines. The engineering drawing of the footings and foundations for a small commercial building appears in Figure 11-1.

In the example, solid foundation walls with continuous footings support the exterior walls. The center columns would bear on the grade beam with concrete piles shown in the center of the plan. Notice that all items of importance are dimensioned

and that such features as piles are dimensioned to their center lines.

The relative simplicity of the example makes a complex marking system unnecessary. In this case, carefully chosen typical sections and adequate notes and call-outs provide the information necessary to create the system.

The desired reinforcing units in the example are called out on the drawing and are shown and labeled on the details and sections. Since the example is uncomplicated, these methods suffice for the reinforcing and eliminate the need for a separate schedule. The addition of a schedule would not be redundant and would serve to ensure more adequately the proper interpretation of the information.

TYPICAL FOOTING AND FOUNDATION DRAWINGS (WITH PILASTERS)

The design drawing illustrated in Figure 11-2 is for a more involved footing and foundation system. This system relies upon continuous foundation walls (actually grade beams) with projecting pilasters and pad footings beneath the pilaster locations. The drawing in this case is based upon solid object lines for the foundation walls and broken lines for the pad footings. The various components in this example are assigned identifying marks.

The foundation walls, as mentioned, serve as grade beams and are marked with the standard designation for beams. The schedule is keyed to the marking system. It provides information as to each beam's size, and the type, size, and number of reinforcing bars in each beam.

The information for the concrete work in the previous example is more complete and detailed when a placing drawing has been prepared for the same project. The placing drawing shown in Figure 11-3 has been drawn for the same building.

Notice that this drawing carefully calls out the exact information about reinforcing where it is to occur. Information is given as

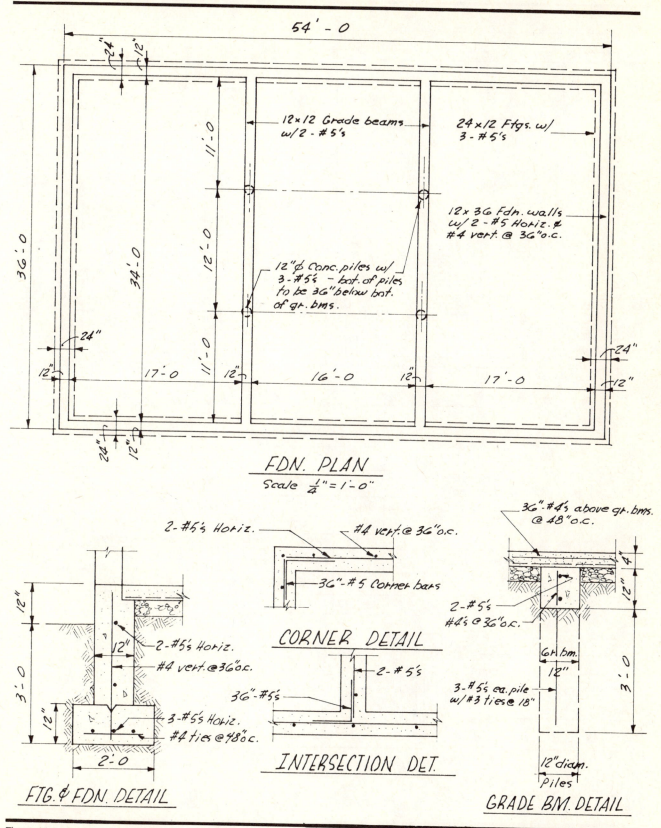

Figure 11·1 Small building foundation and details

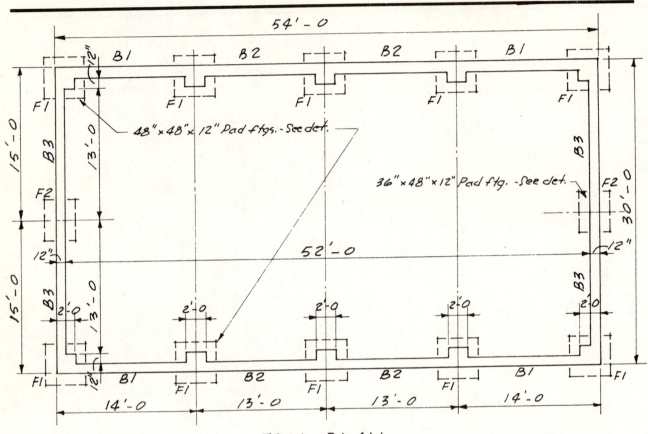

FDN. PLAN

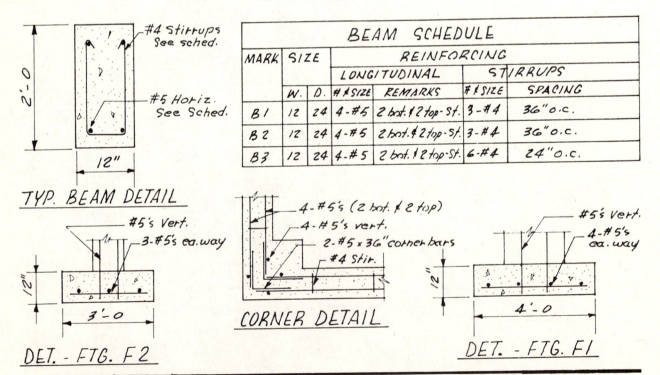

TYP. BEAM DETAIL

		BEAM SCHEDULE					
MARK	SIZE	REINFORCING				STIRRUPS	
		LONGITUDINAL					
	W.	D.	# & SIZE	REMARKS	# & SIZE	SPACING	
B1	12	24	4-#5	2 bot. & 2 top - St.	3-#4	36" o.c.	
B2	12	24	4-#5	2 bot. & 2 top - St.	3-#4	36" o.c.	
B3	12	24	4-#5	2 bot. & 2 top - St.	6-#4	24" o.c.	

DET. - FTG. F2

CORNER DETAIL

DET. - FTG. F1

Figure 11·2 Example of an engineering drawing and schedule

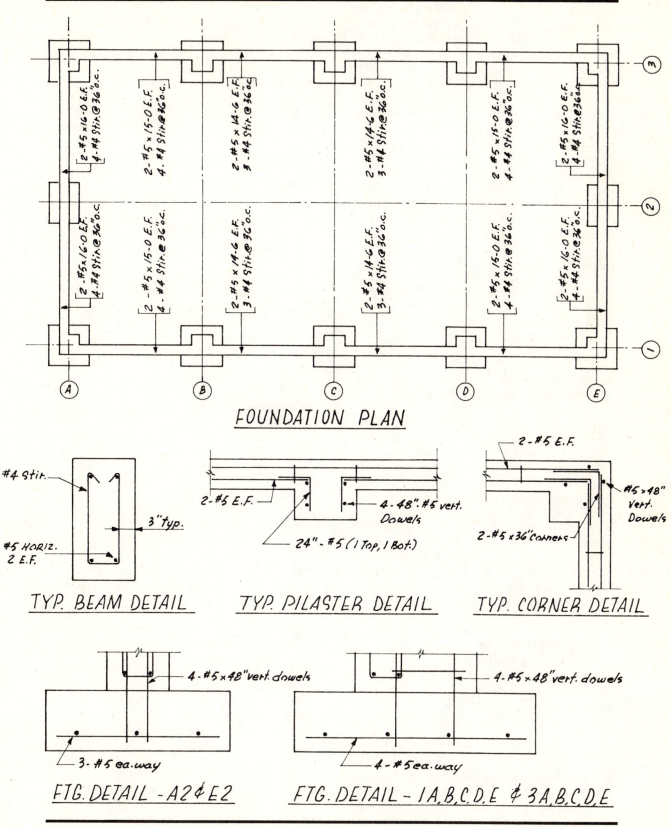

FOUNDATION PLAN

TYP. BEAM DETAIL   TYP. PILASTER DETAIL   TYP. CORNER DETAIL

FTG. DETAIL - A2 & E2   FTG. DETAIL - 1A,B,C,D,E & 3A,B,C,D,E

Figure 11·3 Example of a placing drawing

to whether bars are to be at the top or bottom of members, and next to which face they are to be placed. A numbering system by piers (pilasters) or bays is used on this drawing, rather than the marking system used for the engineering drawing.

Details are provided for the grade beams, footings, and a typical corner. Many designers intentionally maintain a distance of about 3 in. from the face of a member to the reinforcing inside of it. When the note indicates inside face (I.F.), outside face (O.F.), or both faces (E.F.), it is clearly assumed that the 3-in. minimum clearance will be maintained.

Dowels are bars which are placed in one pour to bond a later pour or component to it. In the example, dowels are to be installed to join the actual wall pilasters to the foundation. These dowels appear and are called out on the drawing.

WALL-BEARING STRUCTURAL SYSTEMS

Many of the smaller commercial buildings lend themselves well to the wall-bearing framing system. With the proper choice of beams or panels, adequate spans can be provided to meet many commercial building space requirements.

A number of components can be used to span between load-bearing walls to support floor and roof decks. Beams and decks can be poured in place. The precasting and prestressing industry produces a variety of units well suited for this purpose.

Precast and prestressed roof and floor panels are available in flat, hollow-core, and channel slab forms. Prestressed beams can be obtained separately or combined with decking in the configuration of the channel slab, monowing (F-section), single T, and double T.

The walls for wall-bearing construction have various possibilities for using different materials. Certainly the structural concrete-block wall is widely used for this purpose and may, or may not, be veneered with brick or other finish materials.

Load-bearing walls of solid or composite brick or stone can support wall-bearing structural elements but are often avoided because of increased cost factors.

Poured concrete bearing walls can be used to support floor and roof construction. They are most often limited to basement or below-grade applications. This fact is due, in part, to the difficulties involved in forming above ground and the resulting appearance.

EXAMPLE OF DRAWINGS FOR PRESTRESSED DOUBLE T'S

The drawing in Figure 11-4 represents a framing (or structural) plan for a small commercial building. The single-story structure has load-bearing concrete-block walls with 4-in. brick veneer. The roof system is made up of prestressed double T-units which are supported by the exterior walls.

The drawing indicates the size, direction, and location of all of the prestressed units. Notice that areas necessary for openings in the roof are called out and shown very clearly. This practice permits the simple process of *forming out* the desired openings rather than breaking them out on the job. The panels carry designating marks to ensure their proper fabrication and placement.

Since some of the double T's extend to produce an overhang, it is important to give the governing dimensions for these, as well as for reoccurring panels. The accompanying details show dimensions and erection features necessary for the system.

The precasting company, as mentioned before, will create its own drawings for the prestressed units in accordance with codes and standards. In this example, then, it is unnecessary to show in detail the reinforcing of the double T-units.

Other precast and prestressed components would be drawn in a manner similar to the example for double T's. The information necessary would still include the governing dimensions, any special openings or end

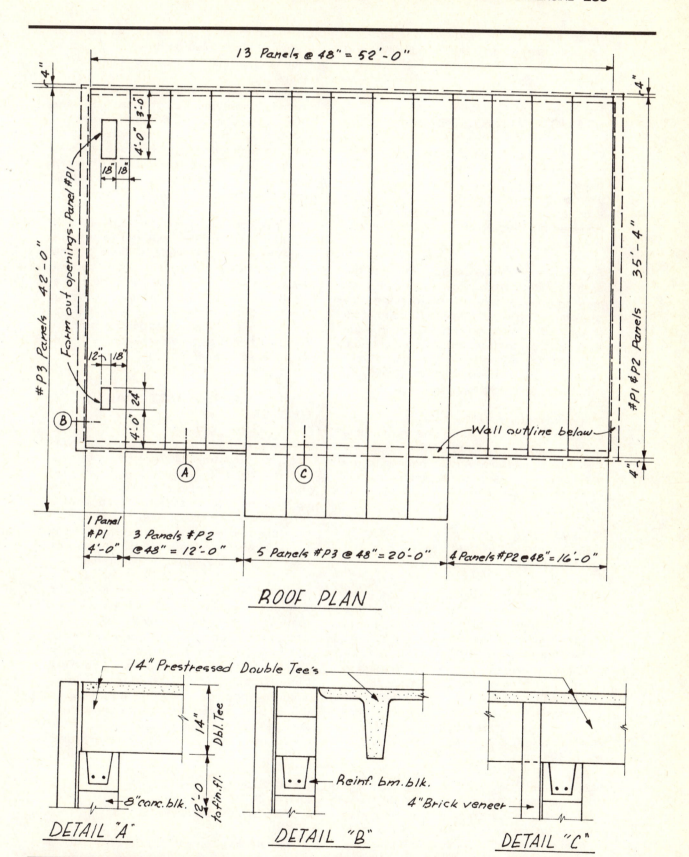

ROOF PLAN

DETAIL "A"   DETAIL "B"   DETAIL "C"

Figure 11·4 Prestressed concrete roof plan

treatments, details showing regular and special bearing and end conditions, and joints.

BEAM-AND-COLUMN STRUCTURAL SYSTEMS

Commercial buildings, particularly those which are multistory, can make excellent use of beam-and-column framing systems. These systems permit relatively rapid erection and can be quickly enclosed with non-load-bearing walls (or curtain-wall panels).

Several floors can be constructed with the system, with much less overall weight than would be necessary for wall-bearing construction. The bays formed by beam-and-column systems can be designed in such a manner as to frequently allow the columns to fall where interior dividing walls occur, thereby keeping them from encroaching on usable interior space.

It is important in detailing beam-and-column systems to assign clearly understandable marks to all components of the structural system. The detailer should adhere to the standard marks and nomenclature to prevent his drawings from being misread.

EXAMPLE OF BEAM-AND-COLUMN DRAWINGS

The partial engineering drawing in Figure 11-5 is for the second (and top) floor of a two-story commercial building using a form of the beam-and-column system. In this example, the columns are round and feature integral column *capitals* (the conical section at the top of the column). The floor slab is thickened in a square area directly above the column capital. The floor-slab reinforcing varies in areas between columns (column strips) and in the center areas of the bays.

Notice that the framing plan contains marks assigned to the component members and information regarding the slab reinforcing bars. The dimensioning system ties down all columns by citing dis-

tances to their center lines. The overall dimensions would also be provided. The slab thickness is called out on the plan itself.

Details would be provided including both specific and typical sections. These details would show the desired reinforcing and the minimum clearances intended for reinforcing in the slab. In this case, the slab reinforcing is to be placed in such a manner that it is no closer than $3/4$ in. to the bottom and top surfaces of the floor slab.

The full drawing would include a beam schedule which called out the reinforcing necessary for the various floor beams. Data regarding the strength of concrete desired and factors governing the placement of reinforcing steel would be covered in the notes contained on the drawing sheet. Details for the columns would be contained on a separate sheet, along with a column schedule.

LONG-SPAN STRUCTURAL SYSTEMS

The advantages of the larger long-span concrete structural elements are best applicable to large commercial structures. The larger sizes of the prestressed units such as the channel slab and the monowing offer long-span capabilities with relative ease and can be provided with end-bearing by load-bearing walls (with pilasters) or by beam-and-column structural systems.

Many other concrete systems exist with long-span attributes. The thin-shell systems of domes, vaults, hyperbolic paraboloids, and conoids offer large areas of open space for commercial purposes. These systems are extremely complex both in design and detailing procedures. They are not, for that reason, included in this volume, since they require experience and understanding far beyond that of the students for whom this book is intended.

EXAMPLE OF DRAWINGS FOR LONG-SPAN SYSTEMS

The example shown in Figure 11-6 is for a

system based upon the prestressed single T-unit. The drawing shown is the roof-framing plan for a larger commercial structure with the accompanying details. The single T-units gain end-bearing from the projecting concrete-block wall pilasters.

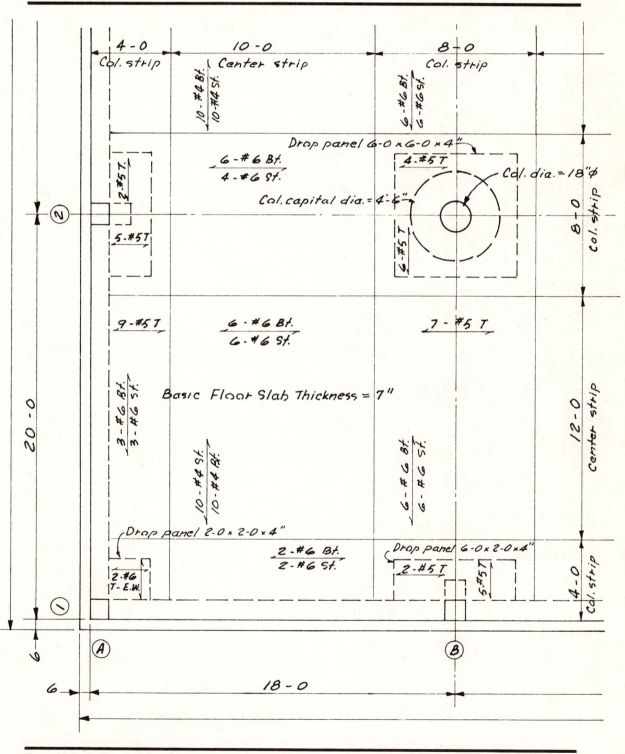

Figure 11·5 Portion of second floor engineering drawing

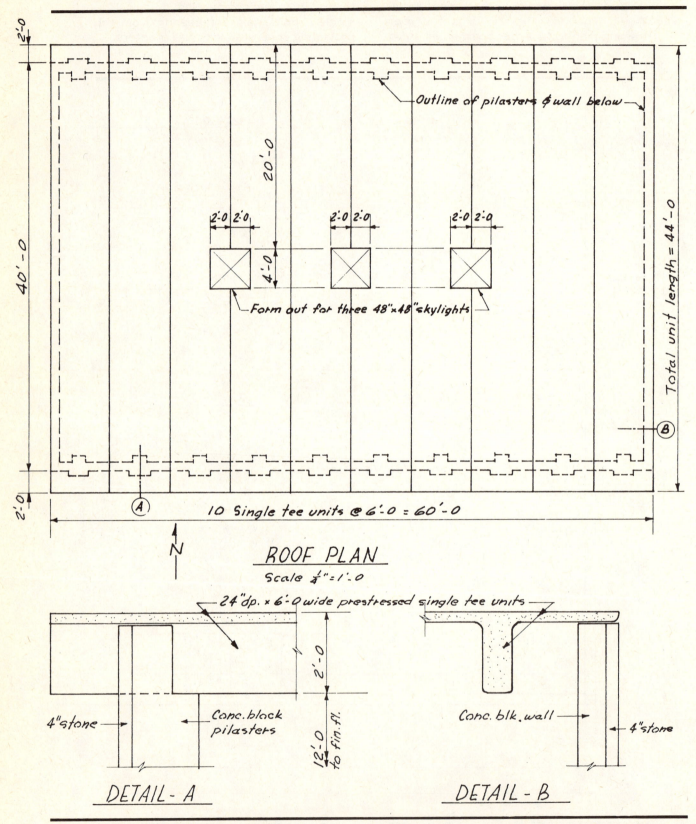

Outline of pilasters & wall below

20'-0

2'-0 2'-0   2'-0 2'-0   2'-0 2'-0

4'-0

Form out for three 48"x48" skylights

40'-0

Total unit length = 44'-0

2'-0

2'-0

10 Single tee units @ 6'-0 = 60'-0

A   B   N

ROOF PLAN
Scale ¼"=1'-0

24"dp. x 6'-0 wide prestressed single tee units

2'-0

12'-0 to fin. fl.

4"stone   Conc. block pilasters

Conc. blk. wall   4"stone

DETAIL-A   DETAIL-B

Figure 11·6 Roof plan for prestressed single T-units

This project uses 8-in. concrete-block exterior walls with 4-in. stone veneer. The resulting wall is 12 in. thick. The pilasters project both inside and outside of the wall line and, like the balance of the wall, are wrapped on the exterior with 4-in. stone veneer. The pilasters serve as an appearance factor. The single T-units are shown on the framing plan. This structure uses the exposed prestressed units as the finished ceiling inside. Skylights provide natural interior lighting, and the roof openings required for the skylights are clearly shown on the framing plan.

The details shown include those explaining the end-bearing, edge-bearing, and attachment of the prestressed units.

SUMMARY

Concrete structural systems can be used to excellent advantage in many commercial structures. The low maintenance costs, satisfactory fire ratings, and appearance of well-finished concrete units make them attractive to the commercial building industry.

Commercial buildings utilize concrete structural members in wall-bearing, beam-and-column, and long-span formats. Each of the three framing types permits the open space so desirable in commercial structures and facilitates the interior remodeling necessary to meet the changing needs of commercial usage.

Footings and foundations of commercial buildings encounter larger and more varied loads than those of smaller building types.

The foundation system is an important part of any commercial structure. Concrete structural members are often relatively heavy, and walls, footings, and columns must be designed accordingly. Joints of the structural system are often exposed in commercial applications and must be designed and detailed with their appearance in mind.

Poured-in-place and precast (prestressed) systems are used for commercial buildings. The amount of drawing necessary varies with the type and complexity of the structural system chosen for the project. Drawings must conform to the standard procedures and practices accepted industry-wide, and they must be accurate and complete.

Problems—Chapter 11

1. The sketch shown is the rough floor plan for a small commercial building. Prepare the engineering drawing for its foundation and footing system based upon the following:
   a. Depth of footings below grade = 36 in.
   b. Finish floor to finish grade = 8 in.
   c. Width of foundation wall = 12 in.
   d. Footings to be 24 in. wide and 21 in. thick
   e. Reinforcing in footings = 3-#5 bars
   f. Reinforcing in foundations = 3-#5 bars horizontal and #5 bars vertically @ 48 in. O.C.
   g. Floor slab thickness = 4 in. with 6 × 6–10 × 10 WWF.
   h. The center grade beam to be 8 × 24 in. with 2-#4 bars horizontal
   (see the diagram on the next page)

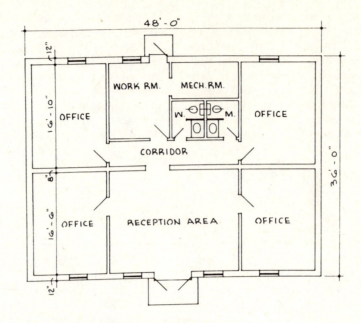

FLOOR PLAN

2. Prepare a placing drawing for the footing and foundation system drawn in problem 1.

3. Prepare a roof-framing plan for the commercial building shown in the illustration. The roof system is to be constructed using 8-in. prestressed double T-units, 48 in. wide and 28 ft long. (Refer to Figure 11-4.) Provide four 12-in.-square openings for skylights in the location shown. Include all details necessary for the roof structure. The load-bearing exterior walls are of 8-in. concrete block with 4-in. brick veneer, and the distance from floor to Bot. of T-units is 9 ft 4 in.

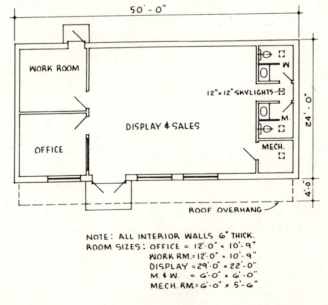

NOTE: ALL INTERIOR WALLS 6" THICK.
ROOM SIZES: OFFICE = 12'-0" x 10'-9"
WORK RM.=12'-0" x 10'-9"
DISPLAY =29'-0" x 22'-0"
M. & W. = 6'-0" x 6'-0"
MECH.RM.= 6'-0" x 5'-6"

FLOOR PLAN

4. Using the same plan shown in problem 3, prepare an engineering drawing for the foundation and footing system. The footings are 30 in. below grade; finish floor to finish grade distance is 6 in.; footings are 24 in. wide and 12 in. thick; and foundation walls are 12 in. wide. Use the same footing and foundation reinforcing specified for problem 1.

After Studying This Chapter You Should Be Able To:

1. Name the main structural framing systems which are most used in industrial building.
2. Cite the features which make concrete a logical choice for many industrial projects.
3. Explain why systems which permit easy remodeling and additions are attractive to industry.
4. Describe the possible effects of traveling industrial equipment on the structural system, and outline steps which can be taken to counteract those effects.
5. Explain how the procedures for detailing industrial structures vary from those used to detail commercial buildings.

INTRODUCTION

The design of structural systems for industrial facilities must account for the factors of heavy loads, possible damage from moving equipment, and the factors of fire and wind damage. Concrete structural systems often have the necessary attributes sought for industrial buildings.

Adequately designed structural systems of concrete are stable, require a minimum of maintenance, and are fire and corrosion resistant. Concrete provides far more sound-deadening than does steel. Steel systems tend to vibrate and amplify sound, whereas concrete furnishes a damping effect in itself.

Wall-bearing, beam-and-column, and long-span framing systems are all used in con-

chapter 12
concrete structural systems- industrial

6. Describe the various types of floor slabs used for industrial buildings.
7. Define *joist-slab* and *waffle-slab* construction systems.
8. Describe some of the various joints and connections used in concrete construction. Compare the connections used for concrete beam-to-column joints with those used in steel construction.
9. Define *chamfer*, and explain the purpose it serves.
10. Describe the lift-slab method of construction, and explain how the drawings for it vary from those used for regular systems.
11. Explain the tilt-up-slab method of wall construction, and name the drawings which are necessary for such a system.

crete industrial structures. Wall-bearing framing is limited in use to smaller facilities because of its limitations in height and the difficulties it presents in later modifying wall openings. The potential ease in remodeling or adding to industrial structures is a significant factor which should be considered in their design.

The beam-and-column and long-span framing systems can provide the large, open spaces demanded for industrial operations. They also permit future modification with relatively few problems. Both systems can be constructed by poured-in-place methods, but the precast and prestressed units are finding increased application.

The appearance of joints and connections should be kept in mind when detailing industrial structures, but it is not the major factor that it is with commercial buildings. Industrial systems should have joints which are simple and functional. Many electrical and mechanical systems in industrial structures are hung from the structural members. Provisions, as mentioned in previous chapters, can be made for attaching such items. The detailer can call out devices which are placed in the members as they are being poured or cast, thus eliminating a great deal of labor after the structural system is complete.

Traveling cranes, forklifts, and trucks frequently are present in industrial structures. These units pose a threat to the structural system in the form of damage by impact. Columns can be protected by providing enlarged bases. Some detailers call for concrete or steel *bumpers* of some type to ensure that traveling equipment will be kept away from the columns.

DRAWING FOR INDUSTRIAL STRUCTURES

The procedures for detailing concrete structural systems for industrial buildings are essentially the same as those used in detailing commercial structures. As mentioned previously, the quantity of details is often reduced when dealing with standard precast and prestressed components. The producing firm in such instances often provides its own production details conforming to industry-wide standards.

The drawings prepared by the producer of precast components are similar to those for poured-in-place units. To produce standardized units, it is still necessary to have drawings providing dimensions of members and size and placement of the required reinforcing.

Industrial structures require the engineering (design) drawings and placing drawings for all elements of the structural system. Included in the drawings would be the foundation system and all floor slabs, in addition to other framing members and units.

FOOTINGS AND FOUNDATIONS

The loads presented by industrial structures make adequate foundation systems a critical necessity. The vibrations set up by stationary and traveling industrial equipment have a decided effect upon the foundation of such structures. The size of many industrial buildings produces heavy dead loads from the materials of the building itself.

The same types of footings and foundations used in commercial structures are found on the industrial scene. Possibly more piers and piles are used in industrial construction because of the varying site conditions encountered in many industrial locations. The footings and foundation system should be adequately covered with engineering drawings and placement drawings.

TYPICAL FOOTING AND FOUNDATION DRAWINGS

The previous chapter contained examples of typical drawings for commercial building foundation systems. The same procedures are followed in preparing the sample engineering drawing in Figure 12-1. The drawing is for a portion of the footing and foundation system for an industrial structure using the beam-and-column framing system.

The footings in the example are of both continuous and pad footing types. All of the critical items are carefully dimensioned and called out.

The foundation placing drawing for the same structure appears in part in Figure 12-2. The only dimensions appearing on this drawing are those pertaining to the reinforcing steel, but the drawing has been prepared at the same scale as the engineering drawing.

The call-outs on the placing drawing cite the number of bars, their size and length, and their location for each area of the foundation system. Footing reinforcement is called out on the plan in the same manner except for the pad footings. The reinforcing for these units, corners, and pilasters would be on the details drawn on the same sheet.

INDUSTRIAL FLOOR SLABS

Floor slabs in industrial structures must withstand the extremely high unit loads imposed by heavy material storage and heavy equipment. As with other components of the structural system, resistance to vibration is also important.

A number of methods are used in providing adequate floor construction. The lowest floor of most structures is ground-supported and can be easily provided with

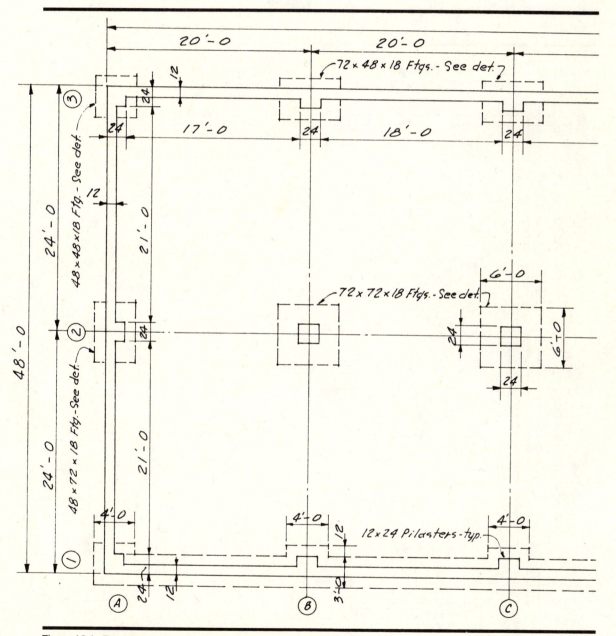

Figure 12·1 Example of a footing and foundation engineering drawing

grade beams and pads to furnish additional support. The upper floors, although supported at increments by beams, columns, etc., must be much more capable of providing their own support.

The prestressed units which combine beams and subdecking are often used, with the addition of a topping or finish slab. Some systems continue to rely upon the solid, reinforced, poured-in-place floor slabs which have been used for years.

A poured-in-place system resulting in a waffle slab is used a great deal.

Waffle slabs The waffle slab is created by use of steel pans (sometimes termed *domes*) to form a grid of concrete joists with an integral slab above them. This system is accomplished in a single pour and results in a continuous system without the bulk and weight of a regular slab. The finished product resembles a waffle when viewed from below.

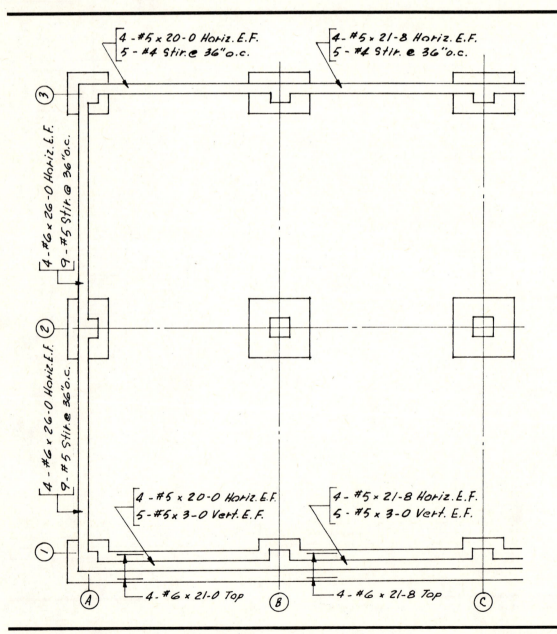

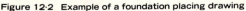

Figure 12·2 Example of a foundation placing drawing

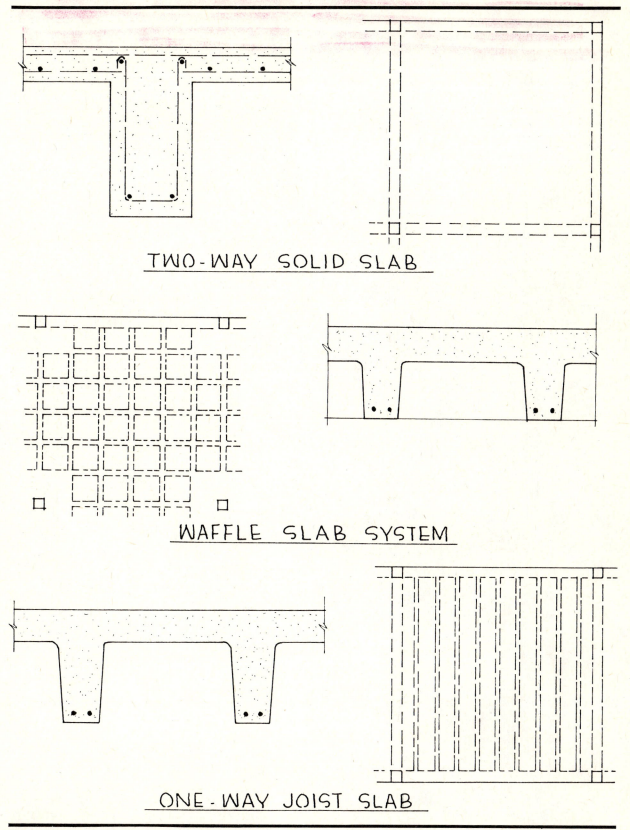

TWO-WAY SOLID SLAB

WAFFLE SLAB SYSTEM

ONE-WAY JOIST SLAB

Figure 12·3 Typical floor systems in plan and section

The form pans are available in a range of sizes and depths. Most dome pans create joists at right angles to each other with the center-to-center spacing of 36 in. Pans are also available in sizes to create 24-in. joist spacings and to fill in nonstandard spaces as required. Depths of the pans vary from 6 to 14 in. in increments of 2 in.

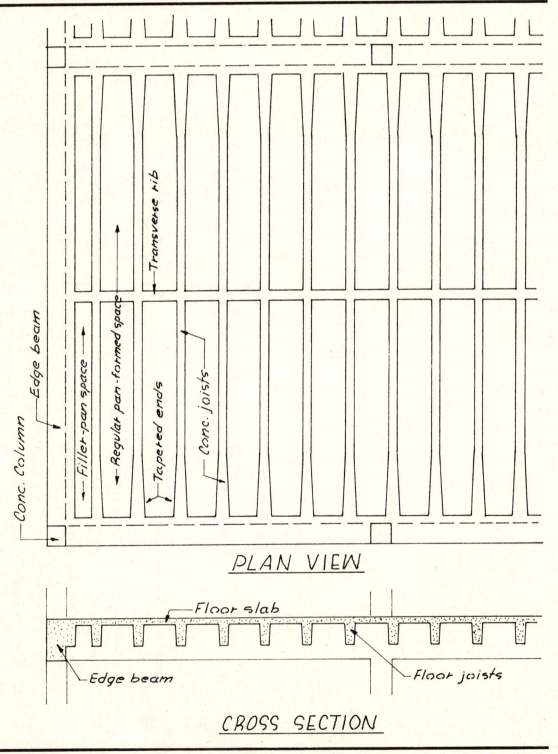

PLAN VIEW

CROSS SECTION

Figure 12-4 Joist-slab plan and section

The domes are generally eliminated in the immediate area of a column (or decreased in the depth of void they form) to create a thicker slab (or *drop panel*) immediately above the column. The joists formed by the pans are reinforced in the manner used for any poured-in-place beam or joist. Top slabs vary with design requirements, but a minimum thickness of 3 in. is usually adhered to.

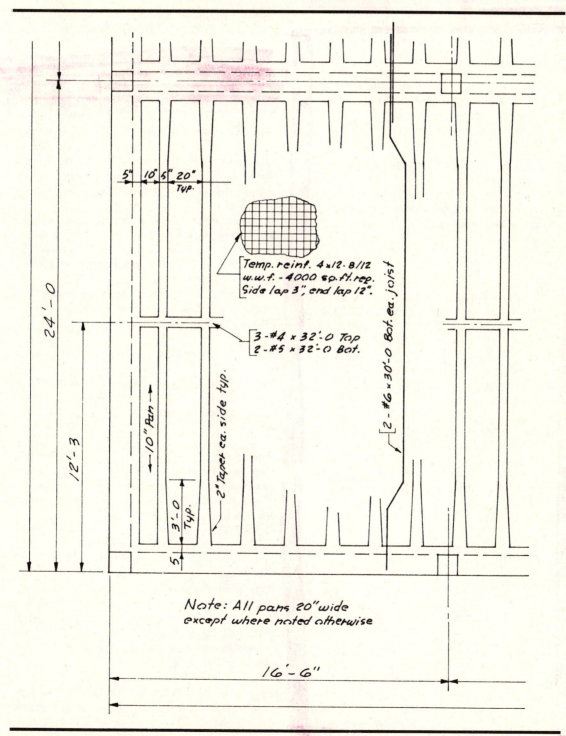

Figure 12·5 Partial placing drawing

Dome forms are available to create rectangular voids, as well as square spacings. Similar units (forming continuous joists in one direction only) are available for one-way joist slabs. The sides of dome forms slope inward from bottom to top. This permits ease in removing them for use on other pours.

Figure 12-3 shows various types of floor-slab construction available for concrete structural systems. Roof slabs can be created in identical manner and are most often of less depth because of the lack of the same live loads and dead loads present on the floors of a building.

Joist slabs Many joist-slab installations provide a taper at the ends of joists as they join with the beams of the structure. This practice creates wider joists at the point where they gain end-bearing and where the stresses are, of course, greatest.

The partial plan view and section shown in Figure 12-4 indicate details of a joist-slab system. The amount of taper created varies with the width of dome form used. The amount of taper shown in the example is 2 in. on each side for 20-in. forms. The length of the taper may vary with the particular design, but in the example it is 3 ft 0 in.

Joist-slab placing drawings Placing drawings for joist-slab structural systems conform to the practices followed in preparing placing drawings for any poured-in-place system. The drawing in Figure 12-5 is a partial placing plan for a joist-slab industrial floor.

The total drawing would contain the usual plan, details, and schedules necessary for any project. Notice that reinforcing specifications are noted for each component at the point where they occur. The tapered ends of the joists are shown in the plan.

The size of pans is noted as applying to all cases except those noted specifically as being of another size. It is common practice to call out the slab reinforcing on the plan itself, as is the case in the example. It is common procedure to indicate reinforcing

for slabs in a separate schedule, giving bar specifications, sizes, and desired bends. Such a schedule does not usually indicate the total number of bars required for the slab.

JOINTS IN CONCRETE INDUSTRIAL STRUCTURES

The joints and connections used for beams, joists, and columns in concrete structural systems resemble those used in structures based upon steel or wood. In industrial work, joints must be capable of withstanding extremely heavy stresses. Junctions of intersecting concrete members may appear different owing to the fact that many are created in a composite pour with the members themselves.

Figure 12-6 shows some of the various joint connections used in concrete which are equivalent to those used in steel. The seated connection used in steel is formed by the column in this case which has integrally formed projections to support the bottom of the beams framing into it as shown in example (A).

Example (B) indicates a column-and-beam junction poured in one pour. In this case, the sloping portions joining beams and column serve in the same way as a framed connection in steel with angle or knee braces.

The column in example (C) has a column capital which collects the various loads and transmits them into the column shaft. The capital serves much the same function as the steel cap plate placed on steel columns. The column capital also provides better end-bearing conditions for the beams or slabs framing into it.

The joint shown in example (D) provides end-bearing assistance for the joist-slab system in the form of a drop panel (full-thickness slab area) above the column. Notice that a column capital is provided in addition to the drop panel.

Example (E) provides a detail of a ledger beam which serves to support the ends of prestressed double T-units. The beam,

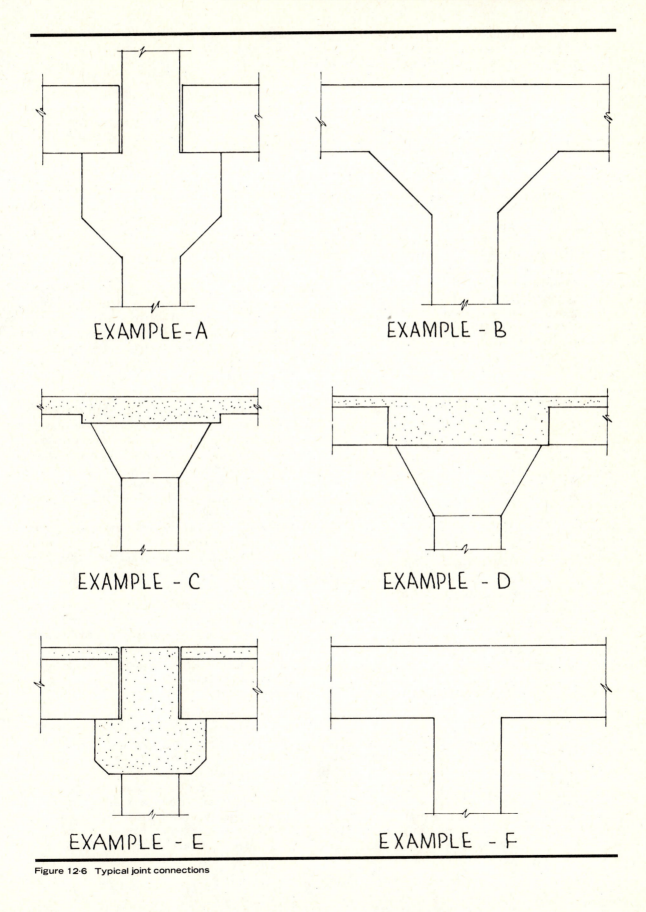

Figure 12·6 Typical joint connections

whether precast or poured in place, features ledges on each side to receive the ends of the prestressed members. Similar joints may be used in poured-in-place projects and to receive the edges of regular floor slabs.

The much-used straight connection of example (*F*) is accomplished in one pour. This type of joint approximates the framed connection used in steel construction and makes no provision for knee bracing or additional connection support.

The shape or configuration of connections in concrete, and member shape to a certain extent, are often the result of having to ensure the removability of formwork. Sloping, tapering, and curved surfaces often permit forms to be removed easily. It is also a means of ensuring that voids in the forms will be more adequately filled with concrete.

CHAMFER OF EDGES

The term *chamfer* refers to the practice of beveling or cutting sharp (90°) corners on a 45° angle. In concrete construction chamfer is often called for on members and connections.

The various sections shown in Figure 12-7 indicate how chamfer is applied to columns and beams. As shown, the proper way of calling out a chamfered edge is to use the word *chamfer* preceded by the distance (along the member's surface) from its beginning to the point which would have been the regular corner. In other words, when the surface of the desired chamfer is considered as the hypotenuse of a 45° by 45° by 90° triangle, the specifying dimension would be that of each of the other two sides (actually the portion being eliminated).

Chamfering serves a very useful purpose. It is extremely difficult to maintain sharp 90° corners and edges on concrete members. The removal of forms often results in edges chipping or crumbling. It is also difficult to assure that concrete extends uniformly into the 90° corners of the forms. Small voids in such instances are

practically unavoidable. Chamfering eliminates much of the problem just cited.

The hard usage of industrial facilities frequently results in damage from collision to the corners and edges of concrete work. The damage is always unsightly, and when large areas of a member are broken away, structural damage can result. Chamfering is one means of holding damage of this nature to a minimum.

EXAMPLES OF LONG-SPAN FRAMING

Industrial structures frequently make use of prestressed members in the long-span framing system. The framing plan for one floor of an industrial structure is shown in Figure 12-8. This building uses the prestressed single T-units which combine beams and deck.

The single T's bear on a poured-in-place concrete beam (girder) and column system. The poured-in-place units are provided with steel connectors which permit welding the steel fasteners of the prestressed units to them. Exterior walls in the example are of concrete block but are not load-bearing owing to the fact that they simply fill in the bays formed by the poured concrete beams and columns. In this instance, the upper walls are supported by the beam-and-column system.

The ends of single T's are often joined for multibays by a field pour which joins them to the stirrups provided in the main beam supporting them. The joining pour may be accomplished at the same time as pouring the topping slab.

When prestressed units span more than one bay at a time, it may be necessary to provide some means of tying them to the intermediate beam or wall below. Stirrups are provided in the supporting member in such cases, and openings are provided in the edges of the prestressed units to permit pouring concrete into the void below. Figure 12-9 indicates two methods for joining units in multibay applications.

EXAMPLES OF BEAMS AND COLUMNS

Industrial structures using the beam-and-column system of framing are detailed in the same manner as such systems explained in the preceding chapters. The main difference in the systems would lie in the size and design of the components for the heavier loads of the industrial function.

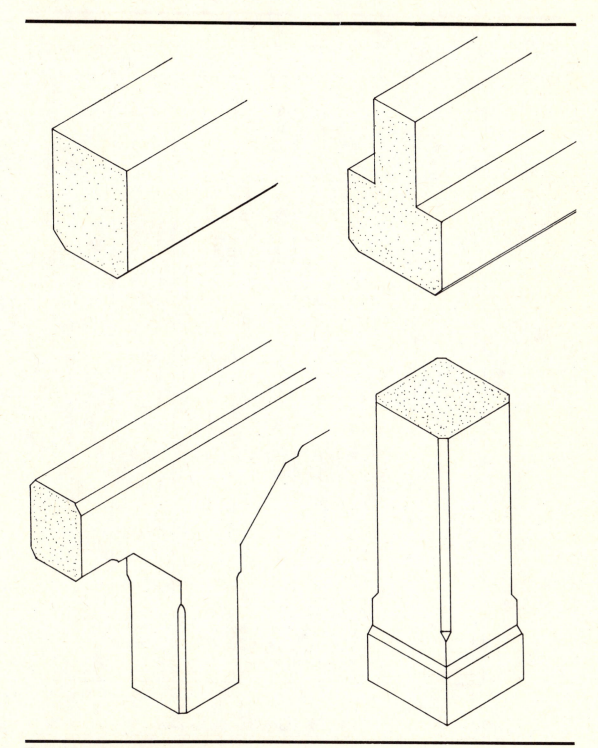

Figure 12·7 Examples of chamfered edges

Industrial buildings often require exterior structures to support conveyors, loading platforms, and drives and walks joining two buildings. The supports for such elements often take a form similar to that of a steel bent (or rigid frame). The support detailed in Figure 12-10 is actually composed of two columns and two beams with the accompanying footings.

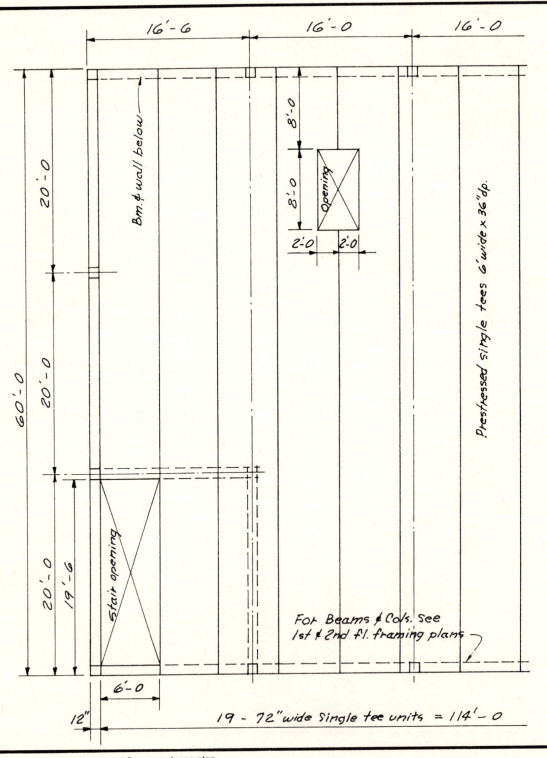

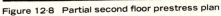

Figure 12·8 Partial second floor prestress plan

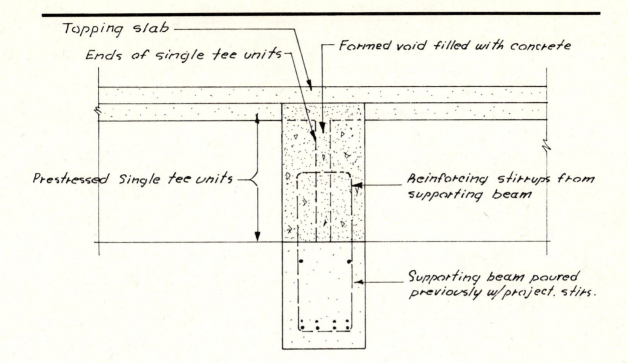

Topping slab

Ends of single tee units

Formed void filled with concrete

Prestressed Single tee units

Reinforcing stirrups from supporting beam

Supporting beam poured previously w/project. stirs.

END BEARING ON PRESTRESSED BEAM

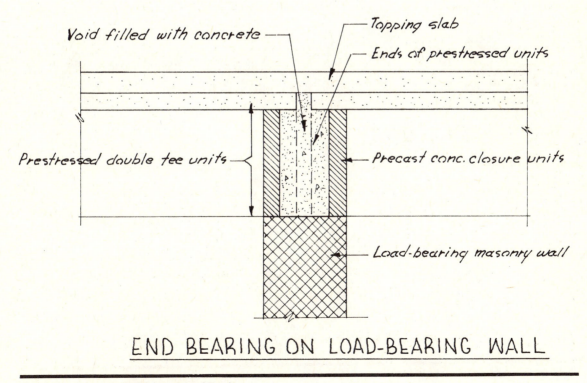

Void filled with concrete

Topping slab

Ends of prestressed units

Prestressed double tee units

Precast conc. closure units

Load-bearing masonry wall

END BEARING ON LOAD-BEARING WALL

Figure 12·9 Methods for joining ends of T-units

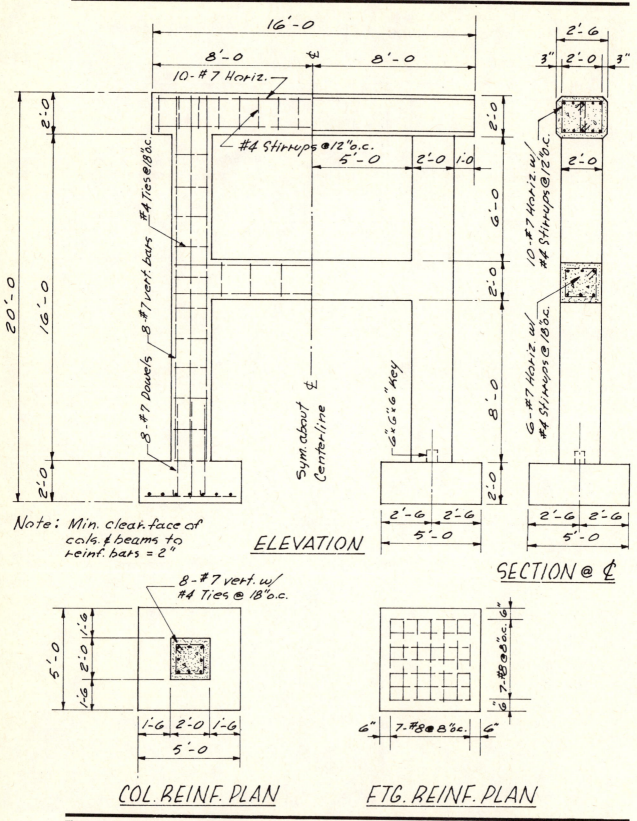

16'-0

8'-0   ℄   8'-0

10-#7 Horiz.

#4 Ties @ 18"o.c.

#4 Stirrups @ 12"o.c.

5'-0   2'-0   1'-0

8-#7 vert. bars

2'-0

8-#7 Dowels

Sym. about ℄ Centerline

6"x6"x6" key

Note: Min. clear. face of cols. & beams to reinf. bars = 2"

20'-0   16'-0   2'-0

2'-6

3"   2'-0   3"

10-#7 Horiz. w/ #4 Stirrups @ 12"o.c.

2'-0

6-#7 Horiz. w/ #4 Stirrups @ 18"o.c.

2'-6   2'-6

5'-0

2'-6   2'-6

5'-0

ELEVATION

SECTION @ ℄

8-#7 vert. w/ #4 Ties @ 18"o.c.

5'-0   1'-6   2'-0   1'-6

1'-6   2'-0   1'-6

5'-0

6"   7-#8 @ 8"o.c.   6"

6"   7-#8 @ 8"o.c.   6"

COL. REINF. PLAN

FTG. REINF. PLAN

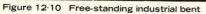

Figure 12·10 Free-standing industrial bent

The intention of the detailer in this case is to pour the footings with extended dowels to receive the pour for the balance of the structure. The columns and beams would be poured in a single composite operation.

The unit shown in the example features chamfered edges on the top beam. The footings are of the pad footing variety and are heavily reinforced with bars in each direction. In addition to relying upon dowels to join columns to footings, the drawings also call for the joint to be keyed as well.

The keying method used here consists of forming a cubical projection in the center of the footing at the time it is being poured. When the column is poured it would, of course, surround and cover the projection, thereby locking the footing and column together. The purpose in providing the key is to assure that the column will not shift horizontally on top of the footing.

LIFT-SLAB CONSTRUCTION

Industrial structures (and commercial structures) sometimes rely upon structural systems of the lift-slab type. The methods of preparing engineering and placing drawings conform to those used for more common systems, since it is basically the erection method which is different.

Many lift-slab projects use steel columns, although concrete columns are also used. The foundation system is created in the regular manner, and the columns are erected on (and fastened to) the foundation. The ground floor (or base) slab is then poured over the foundation and around the columns.

A layer of bond-preventing material is placed over the top of the ground slab, and *lifting rings* (steel collars) are placed around the columns. Another floor slab is poured on top of the first after the previous operations have been accomplished. The entire procedure is repeated for as many floors as are required, with a final slab poured for the roof if desired.

The slabs are then lifted (actually sliding up the columns) to the necessary height

(elevation). When the slab is in position, it is secured to each column by welding, pinning, or bolting the lifting rings to the columns. The operation is repeated until all slabs are permanently secured in place. Lift slabs may be prestressed for added strength.

Slabs formed for floors in lift-slab structures are generally of the flat-slab type. Coffered or waffle slabs are used occasionally where heavy loads or extremely long spans are required. Slabs must be allowed to reach adequate compressive strength before being lifted.

Many designers require that actual compressive strength be established as being 2,700 psi (through standard cylinder tests) and that the slab be a minimum of 14 days old prior to lifting.

Building layouts must be extremely simple, with reoccurring bay spacings, if the efficiencies of lift-slab methods are to be realized to their fullest extent. Such items as vertical chases, stairwells, and elevators are best placed on the ends or completely outside of the slab line of the building.

TILT-UP-SLAB SYSTEMS

Some concrete structures use a tilt-up system to create walls for the building. In this method of construction, concrete wall panels (slabs) are cast in a horizontal position on the ground adjacent to (or on the floor of) the structure. Once properly set and cured, they are tilted into their vertical position and fastened permanently to the structural system.

The method has been used for multistory structures but is most advantageous when confined to one or two stories. In the case of taller structures, the walls for one story are placed and secured, the floor above is constructed, and the process is repeated.

Provisions must be made in casting the wall slabs for lifting and attachment hardware and devices. The reinforcing of the tilt-up slabs is essentially the same as that for any concrete wall slab, except that it

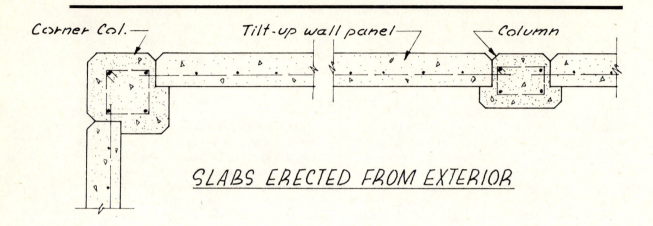

Corner Col.   Tilt-up wall panel   Column

SLABS ERECTED FROM EXTERIOR

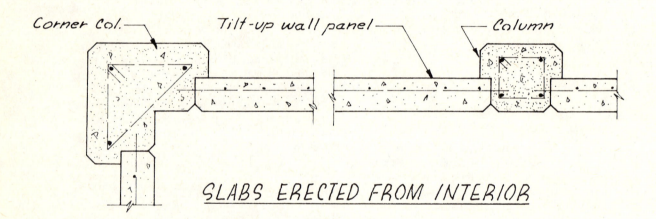

Corner Col.   Tilt-up wall panel   Column

SLABS ERECTED FROM INTERIOR

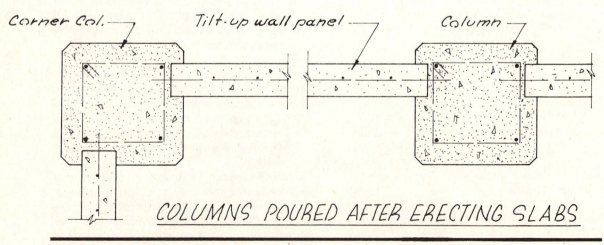

Corner Col.   Tilt-up wall panel   Column

COLUMNS POURED AFTER ERECTING SLABS

Figure 12·11   Three methods for tilt-up slabs

must be adequate to withstand the varied additional stresses imposed by the lifting and tilting-up process.

The drawings for a tilt-up-slab structure resemble those for any other regular system, except that details must be provided for the wall panels and their attachment. Figure 12-11 illustrates in plan some typical joints for tilt-up walls at the columns.

Some tilt-up systems place the wall slabs after the columns have been erected. Other systems pour the columns after the wall panels are in place. The latter system permits the columns to overlap the panels, which results in far more attractive and more weathertight joints.

Tilt-up wall panels can be cast in such a manner that they have prefinished colored and textured surfaces. Depending upon the design, the slabs may be cast with recessed areas for appearance factors. It is also possible to cast panels which contain insulating material or which have one face composed of lightweight insulating concrete.

SUMMARY

A number of methods of construction have been presented in this chapter which relate to the structural systems of concrete industrial buildings. Concrete can provide excellent structural systems which have a number of attributes demanded in industrial facilities.

Concrete systems of proper design provide satisfactory fire ratings and resist damage by many of the corrosive elements present in industrial operations. The vibrations and noise set up within the structural system of industrial buildings by equipment can be held to a minimum in concrete structures.

Small industrial structures use the wall-bearing framing system frequently, but the larger facilities are most often based upon the beam-and-column and long-span systems. These two systems permit much more ease in future remodeling and in later plant additions.

Poured-in-place, precast, and prestressed concrete structural members are all found in industrial construction. The drawings necessary for industrial structures resemble those necessary for commercial buildings and encompass both engineering (design) and placement drawings. All drawings should conform to the format and practices accepted throughout the industry.

Tilt-up-slab construction and lift-slab construction are both used to create industrial structures. The detailing process does not vary appreciably from that used for the regular flat and waffle slab systems. The main difference lies in the erection procedure and the details necessary to facilitate it.

The fact remains that, large or small, poured in place or precast, any structural concrete building can be (and should be) clearly and precisely detailed. The accurate, conscientious detailer, following the practices and standards of the industry, can communicate exactly the criteria and details for erecting the structure. His work can save immense amounts of time, labor, and money during the course of the actual construction of the building. It is his responsibility to do just that.

Problems—Chapter 12

1. The sketch in the illustration is the partial floor plan of an industrial building. Prepare the engineering drawing for its foundation and footing system. The following specifications apply:

    a. Depth of footings below grade = 60 in.
    b. Finish floor to grade = 10 in.
    c. Width of foundation wall = 12 in.
    d. Footings are to be 30 in. wide and 16 in. thick
    e. Reinforcing in footings = 4-#5 bars horizontally with #4 crossbars @ 48 in. O.C.
    f. Reinforcing in foundation = 4-#5 bars horizontally E.F. and #5 bars vertically @ 48 in. O.C.
    g. Floor slab = 6 in. thick with 6 × 6–8/8 W.W.F.

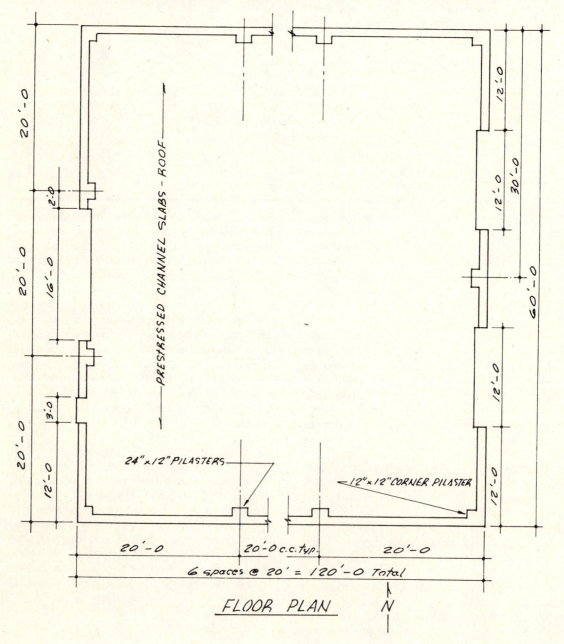

FLOOR PLAN

2. Prepare the foundation placing drawing for the building in problem 1.
3. Prepare the framing plan for the roof of the industrial building shown in problem 1. The system is to be made up of prestressed channel slabs (3 ft 4 in. wide × 18 in. deep) supported by 12-in. concrete-block load-bearing walls.
4. The poured concrete unit in the illustration is to support heavy conveyor equipment for an industrial complex. Prepare detail drawings for the unit. Refer to the example shown in Figure 12-10 of the text for the basic reinforcing required.

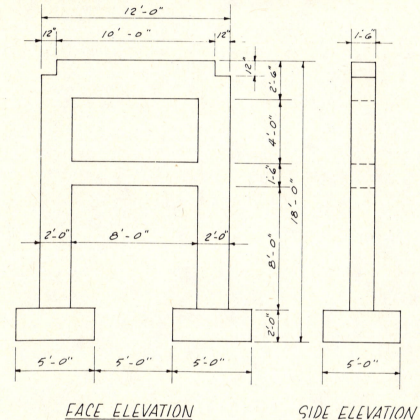

FACE ELEVATION   SIDE ELEVATION

unit four

Courtesy of Marsha Cohen

emerging
technology

INTRODUCTION

Man has searched through the ages for the ideal environments for work, play, and residence. The structures he has created in his attempts to provide shelter for his activities have been varied and unique. Some experiments have ended in dismal failure, while others have led the way to an increasingly sophisticated technology of construction.

Two very important factors must be present, if attempts to produce innovative structural systems are to be successful and capable of contributing to the future generations of structures. The first factor, that of the spark of creative genius, must be present to launch the quest for improved technology into the realm of the unknown and untried. The second neces-

the newer systems mentioned herein are fast approaching the necessary degree of acceptance and standardization.

The material which follows is provided to acquaint the student with the types of structures he may be called upon to detail in the future. Since these systems are encountered much less frequently today than are the more established structural systems, no attempt has been made to describe the detailing procedures necessary for their construction.

INFLATABLE STRUCTURES

The use of inflatable structures is increasing at a rapid rate. These structures combine the structural system with the enclosing function to provide quickly erected

chapter 13
emerging technology— an overview

sary ingredient is the logical, painstaking, scientific and engineering inquiry and documentation which can assure that the flight into the unknown will not end in utter catastrophe.

Any experiment carries with it the element of risk. The magnitude of possible error can be decreased if it is a carefully calculated risk. By building upon the known principles and theories, the designer can reasonably assure the expected outcome of structural experiments.

A number of new structural systems and types have been proved successful and are now in the process of gaining wider acceptance. Along with securing acceptance by the construction industry and the public, the new systems must be standardized if they are to be widely used. Some of

shelters with excellent capabilities for temporary, semipermanent, and seasonal use.

The system uses reinforced fabric to form (in essence) a balloon. The structure is inflated and maintained in its erected position by pressurizing its interior. Fans serve as pumps to maintain the necessary air pressures within the structure. Entrances are most often double entrances which serve as air locks.

Spans available with inflatable structures are such that the structures can be used for gymnasiums, display areas, storage areas, and for some manufacturing processes. Some commercial firms have used the structures to house their offices and operations while more permanent facilities were under construction.

The shape of most inflatable structures in use currently resembles that of the Quonset buildings so popular in World War II days. In cross section, the structure conforms to the radial arch form and functions much the same as a continuous barrel vault. The shape is effective in absorbing stresses and cuts exterior wind loads to a desirable minimum.

As the advance of technology creates fabrics with greater strength and durability, the inflatable structure will continue to approach more closely the degree of permanence associated with the rigid structural systems so widely used today. It already serves as a solution to enclosure problems in inaccessible areas and as an extremely rapid method of providing shelter for people, processes, activities, and materials.

SPRAYED OR FOAMED STRUCTURES

Structures are currently being built based upon a sprayed or foamed system. Both systems require a form or mold, over which the structure is created. The result is a structure which is actually a shell.

Forms for these structures may be built up from other rigid materials, or they may be an inflatable form similar in concept to the inflatable structures discussed previously. The forms are often removed after the structure is self-supporting. Some systems use insulating material for the form and leave it in place as a permanent part of the structure.

The material for these buildings may be lightweight concrete. The concrete is actually sprayed (or blown) onto the form. Once it has hardened, the concrete forms a rigid, self-supporting shell. Reinforcing is placed, wherever necessary, prior to the spraying process.

Plastic materials are available which foam to form a cellular lightweight material. The material resembles that used to fill hollow walls of thermal-insulated containers, refrigerators, boats, etc. These materials can be applied to a structural form to create a rigid shell similar to that created

with sprayed concrete. The foamed plastic material has excellent insulating qualities but must be coated with materials capable of withstanding the elements.

The shape of most sprayed or foamed structures is based upon curved surfaces. Dome shapes, vault shapes, and egg shapes are frequently found. The structural adequacy of the buildings is greatly increased by avoiding straight vertical and horizontal surfaces. They must rely, after all, upon the shell itself to transmit all stresses to the ground.

The construction methods employed in the structures can be adapted to both large and small buildings. The availability of forms is a limiting factor. Shells, of any material, have a tendency to crack from thermal expansion and contraction. This factor is another deterrent to the present use of these structures for enclosure of extremely large areas.

SLIDE-IN MODULE CONSTRUCTION

A system is being used which combines an old structural concept with a new one. This particular system relies upon a beam-and-column structural system which is completed without floors or walls. The structural system is most often of concrete or steel.

Finished rooms (or areas) are constructed separately as self-contained boxes (or modules). These units are lifted to the proper height so that they can slide into the bays of the structure in the same fashion as drawers in a desk or cabinet.

The system permits the construction of the modules (rooms) to take place in an enclosed area, where mass-production techniques can be employed to save material, time, and labor. These structures can be erected quickly and efficiently.

A takeoff from the slide-in system (but similar in concept) is one based upon stacking prebuilt modular units. In this format, the rooms are created and finished in similar fashion, but they do possess structural capabilities. The finished modules are then stacked on top of each other and fastened to each other to form a rigid structure.

The heights feasible with the modular stacking system are considerably less than those of the slide-in system. This is due, of course, to the fact that lower modules must be capable of supporting the weight of those above. There is a point at which it is not economical to create modules of sufficient structural capability to carry the immense loads exerted by the modules stacked high above. These systems are best suited to buildings built to provide multifamily dwellings and offices, since the prefinished modules must be of relatively small dimension.

SUSPENSION SYSTEMS

Considerable experimentation has occurred in recent years with various types of suspension structures. Most suspension designs rely upon a central mass for counteracting all compressive stresses and a series of cables for supporting floors, etc. The cables encounter tension stresses rather than the compressive stresses present in the columns of more conventional systems.

Steel, aluminum, or fiber (in the form of cable or rope) is used in various suspension schemes. Most materials show a great increase in allowable stresses when formed into wire cable or rope strands. Steel, for example, increases from the regular structural steel allowables of 30,000 to 40,000 psi to as much as a staggering 160,000 psi when drawn into wire and cables. This factor would seem to indicate that suspension structures should be economical in terms of material amounts required. By like token, the structural elements (cables) could be of very small cross-sectional dimension.

The materials used with the cables or ropes of such structures vary. Concrete, wood, steel, and fabric have all been used in some fashion in the construction of suspension buildings. The fireproofing of the suspension cables is often an important

factor, and when done properly it increases their size appreciably.

Under actual conditions, the various components of suspension structures are subject to vertical movement, since loads tend to shorten the main compressive unit while elongating the cables. The main mass must also resist most of the wind stresses for the entire structure. Floors have a tendency to swing and vibrate. Temperature changes can create serious problems in suspension structures.

The theory of suspension structures has captured the imagination of numerous designers. It carries with it the illusive promise of large areas spanned with an extremely light and delicate structural system. Most efforts to date have been experiments performed in the one-of-a-kind context. The single most complex problem confronting the designer is that of successfully coping with the inherent movement of the structure and its components. Whether a hanging roof for a single-story structure, or an entire multistory building, control of the movement of the system is essential to its success.

THIN-SHELL STRUCTURES

A broad variety of structures have been built using the thin-shell concrete system. They derive their strength in the same manner as an eggshell. Any direct loads and stresses applied to the shell are actually transferred throughout the entire shell, thus distributing them as equally as possible.

Thin shells are created with concrete poured over forms. The reinforcing steel becomes extremely important. Since the shells are exposed to view, greater care must be exercised in assuring high-quality finished surfaces. This structural system can provide for large, open floor space and easily permits the use of skylights.

Much work has been done in thin-shell format, but it has not approached the degree of standardization achieved with more regular structural types. Many engineers are unfamiliar with the complicated design procedures required, and many contractors are unfamiliar with and un-equipped to handle the problems present in this type of construction.

Vaulted, domed, and conical shapes are frequently seen. The entire area of thin-shell construction is growing rapidly and is destined soon to become one of the regular structural system types. Economy of material and labor are certainly factors in the success of the system.

SUMMARY

This unit has presented short descriptions of exciting recent developments in the area of structural systems design. These systems, along with the space frames and dome systems mentioned earlier in the text, are destined to have tremendous impact upon our future buildings and their method of construction.

As these structural systems evolve, and new ones emerge, the structural detailer will continue to play an important role in the construction process. Whether detailing the common, time-honored beam-and-column system, or some yet unthought-of structural system, the same rules will apply to the detailer's work. He will then, as now, be called upon to create accurate, complete, and concise structural drawings while conforming to the accepted practices of the industry and the time.